T0180565

Designing Training and Instructional Programs for Older Adults

Human Factors & Aging Series

Series Editors
Wendy A. Rogers and Arthur D. Fisk
School of Psychology
Georgia Institute of Technology – Atlanta, Georgia

Published Titles

Human Factors
& Aging Series

Designing Training and Instructional Programs for Older Adults

Sara J. Czaja
Joseph Sharit

CRC Press
Taylor & Francis Group
Boca Raton London New York

CRC Press is an imprint of the
Taylor & Francis Group, an **informa** business

CRC Press
Taylor & Francis Group
6000 Broken Sound Parkway NW, Suite 300
Boca Raton, FL 33487-2742

© 2013 by Taylor & Francis Group, LLC
CRC Press is an imprint of Taylor & Francis Group, an Informa business

No claim to original U.S. Government works

Version Date: 2012907

ISBN-13: 978-1-4398-4787-9 (pbk)

Visit the Taylor & Francis Web site at
http://www.taylorandfrancis.com

and the CRC Press Web site at
http://www.crcpress.com

Contents

Contents

Preface

Our goal in writing this book was to draw on an understanding of today's older adults—including their demographics, their needs, the challenges facing them, and a realistic appraisal of their abilities and limitations—as a basis for how current knowledge about training and instructional design should be shaped and applied to best accommodate this population of learners. With rapidly emerging technologies, including those in the domains of health management and the work sector, as well as the many products that are continually pervading the consumer market that are capable of positively affecting the quality of life of many older adults, understanding the barriers that older people may face in learning to use these resources is imperative.

The literature on training and instructional design is not only extensive but has a relatively long history. Many of the developments in these areas were driven by the desire to improve learning in our educational institutions and the training of our military and industrial personnel. One of the key challenges in writing this book was to distill knowledge from these vast areas that would be relevant to the problem of training and instructional design for older adults. This was necessary in part because there is no theory of training and instruction that is directed solely toward this population of learners. There are stretches throughout this book where it may seem that the older learner has been neglected as we examine in some depth topics fundamental to or related to training and instruction. However, we have tried as much as possible to ensure that these discussions always relate back to the primary subject of this book, the older learner.

We should also comment about the title of this book as, in principle, there is a distinction between training and instruction. The former is usually associated with more formal programs directed at ensuring that people acquire specific skills, whereas the latter has a more general connotation that can encompass household devices and personal or classroom learning. At times throughout the book the context may make it apparent that the discussion is likely addressing training rather than instruction; at other times, it may be the other way around. For the most part, the terms "training" and "instruction" are used interchangeably. This is because when the focus is on the more general concept of "learning something" these distinctions become blurred, as training and instruction both constitute ways of imparting knowledge and skills to people; thus, the more underlying issue is to understand how learning occurs. With older adults the challenge then becomes understanding how their various limitations (which could arise from normal age-related considerations such as changes in cognitive abilities and motivation) as well as their strengths (such as those that may derive from their lifelong experiences) can affect their dispositions toward and capabilities for learning. Also, as noted or implied throughout the book, designing training and instructional programs for older adults can be especially challenging, as older adults are very heterogeneous and vary tremendously in terms of their backgrounds, skills, knowledge, and abilities.

This book was written with a number of audiences in mind. Given the various consequences that the design of training and instructional programs can have for older adults, this book was written with designers in mind. Each chapter begins with the assumption that the reader does not have extensive knowledge about the subject matter contained therein, and concise recommendations are provided at the conclusion of many of the chapters that can have direct implications for the design of instructional programs and for those individuals who are responsible for the training and performance of older people. However, we feel this book will also be of value to academics, including students at both the undergraduate and graduate levels, who have an interest in the areas of aging, cognition, instruction, and performance, as they may have the opportunity to see these topics tackled from a different perspective. Although the book is not littered with references to scientific works as one usually encounters in the more formal academic literature, a number of references are included and compiled at the end of the book. In addition, works that were not explicitly referred to are listed in the recommended reading sections at the end of each chapter. Overall, we strove to produce a book that would prove readable and enlightening across a wide spectrum of readers.

Finally, many of our insights in this area have been culled from our experiences at CREATE, the Center for Research and Education on Aging and Technology Enhancement, which is sponsored by the National Institutes of Health (National Institute on Aging) and comprises a multidisciplinary and multisite center based at the University of Miami's Miller School of Medicine. CREATE addresses older adult interactions with various forms of technology in the domains of health, work, and the home, and has provided a wealth of opportunities for investigating issues related to training and instruction of older adults. We extend our deepest appreciation to all the researchers, students, and staff who have been involved in CREATE as they have contributed immensely to our ability to write this book.

About the authors

Sara J. Czaja, PhD, received her PhD from the State University of New York at Buffalo. She is a Leonard M. Miller Professor in the Department of Psychiatry and Behavioral Sciences, and a Professor of Industrial Engineering at the University of Miami. She is also the scientific director of the Center on Aging at the University of Miami and the director of the Center on Research and Education for Aging and Technology Enhancement (CREATE). CREATE is a multisite center funded by the National Institute on Aging/National Institutes of Health that involves the University of Miami, Georgia Institute of Technology, and Florida State University. The focus of CREATE is on making technology more accessible, useful, and usable for older adults. Dr. Czaja has extensive experience in aging research and a long commitment to developing strategies to improve the quality of life for older adults. Her research interests include aging and cognition, aging and healthcare access, family caregiving, aging and technology, and functional assessment. She has published extensively in these areas. She is a fellow of the American Psychological Association, the Human Factors and Ergonomics Society, and the Gerontological Society of America. She recently served as a member of the National Research Council/National Academy of Sciences Committee on Human Factors and Home Health Care. She is currently a member of the Board on Human Systems Integration for the National Research Council/National Academy of Sciences.

Joseph Sharit, PhD, received his MS and PhD degrees from the School of Industrial Engineering at Purdue University, specializing in human factors engineering. He is currently a research professor in the Department of Industrial Engineering at the University of Miami, and holds secondary appointments in the Department of Psychiatry and Behavioral Science and the Department of Anesthesiology at the University of Miami's Miller School of Medicine. He is one of the principal investigators in the Center on Research and Education for Aging and Technology Enhancement (CREATE), a multidisciplinary, collaborative center dedicated to issues related to aging and technology use across a variety of domain applications. He is also a research scientist with the Miami Veterans Administration Medical Center, where he is affiliated with the GRECC (Geriatric Research Education and Clinical Center) Laboratory for E-Learning and Multimedia Research. His research interests focus on human–machine interaction, including the assessment of older adult interaction with various technologies. This research is strongly influenced by models of human information processing and the use of human factors approaches such as task analysis and simulation methods as a basis for assessing performance with the aim of developing training and design interventions that can enhance older adult task performance and decision making. His research also addresses human error and system safety and the development of training systems.

chapter one

Introduction and overview

Emerging demographic and societal trends underscore the importance of the topic of training and instructional design for older adults. These trends include: the aging of the population, rapid developments in technology and the diffusion of technology into most settings, the migration of healthcare practices from professional facilities to the home, changes in work processes and organizations, and changes in instructional practices and technologies. Generally, people of all ages, including older adults, must engage in what is commonly referred to as *life-long learning* in order to use new products and devices and perform tasks at work (e.g., use software applications, job procedures), at home (e.g., medical devices, communication and entertainment products), in service environments (e.g., automatic teller machines, self-service ticket kiosks at airports), and in instructional environments (e.g., online learning programs).

Most people rely on some form of training or instruction when they need to learn something new or brush up on a previously learned skill. This training may occur formally through attendance at a training program or individual training sessions, the use of online software, an instruction manual, or informally with help from a colleague, family member, or friend. The overall goal of engaging in these activities is to ensure that the learners or trainees have a meaningful learning experience such that they can use the material successfully at a later point in time or transfer what they learned to a new situation. Generally, effective training and instructional programs lead to competence in a particular skill area as well as higher motivation, learner satisfaction, and increased feelings of accomplishment. In the workplace, good training design has also been linked to increased productivity, fewer errors, and improvements in safety.

Designing training and instructional programs to promote meaningful learning has been a long-standing challenge, especially for older adults. As discussed in the next chapter, older adults are a very heterogeneous group on a multitude of dimensions including skills and abilities, prior learning experiences, and motivation. Many older people also experience anxiety in new learning situations or have a lack of confidence about their ability to learn something new. However, this does not imply that they cannot learn new skills. Data from our research examining technology-based tasks, and from many others, presents a fairly positive picture in terms of the continued learning potential of older adults. Older people may take longer or need more feedback or training support than younger people, however, they are able to learn (Chapter 3).

Unfortunately, because of prevailing stereotypes about aging (you can't teach an old dog new tricks), older adults are often bypassed with respect to training opportunities or are offered training programs that are not tailored to meet their preferences and needs. The goals of this book are to present a state-of-the-science summary of the topic of training and instructional design for older adults and, where possible, to present some basic principles and guidelines regarding best practices for designing training and instructional programs for older people. We begin by discussing emerging demographic and societal trends that are relevant to aging, learning, and training, and substantiate the fact that lifelong learning is an important educational reality.

1.1 Demographic trends

One of the most important demographic trends challenging the government, businesses, healthcare, families, and society is the aging of the population. In 2009, people aged 65 and over in the United States accounted for about 13% of the population and are estimated to represent about 20% of the population by 2030 (Figure 1.1). The older population itself is also getting older. The number of people aged 85+ years (the

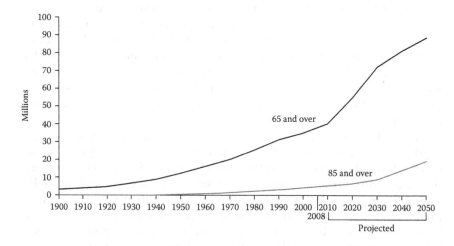

Figure 1.1 Percentage U.S. population age 65 and over. (From Federal Interagency Forum on Aging-Related Statistics, 2010.)

oldest old) was about 5.7 million in 2009 and is expected to increase to about 6.6 million by 2020 (Administration on Aging, 2011). The number of centenarians is also increasing. These trends are paralleled worldwide. In 2006, almost 500 million people worldwide were 65 and older, and that number will increase to about 1 billion by 2030 (National Institute on Aging, 2007). As the population ages so does the population of older adults who need or desire to participate in some form of training or instructional program. Generally, an older learner refers to someone aged 65 or older; however, in some settings such as the workplace, it can refer to people aged 55 and older.

Aging is typically associated with changes in cognition, perception, mobility, and health status. These changes are more pronounced among the oldest old. However, as we discuss in Chapter 2, the population of older adults is very diverse and older adults vary tremendously in educational and health status, literacy, culture/ethnicity, skills and abilities, and of course life experiences. The majority of older adults are also more active than those in previous generations and live independently for longer periods of time. They are also highly motivated to remain productive and engage in learning new skills or new hobbies. For example, recent surveys (e.g., AARP, 2010) indicated that a large majority of older people would like to or need to remain in the workforce on a full or part-time basis. Many are also beginning second careers as they face retirement from their primary occupations. Both scenarios require investments in job and career training programs. In addition, many seniors participate in educational programs to pursue personal interests such as learning a new language, cooking, basic computer or Internet skills, or a multitude of other topics.

All of these issues have tremendous implications for the design of training and instructional programs. For example, age-related changes in cognition have implications for the design of practice protocols and pacing of training (Chapter 3). The demographic diversity of the population has implications for structuring of group learning activities, as do differences in skill levels and literacy. Many of the current generation of older adults also have limited technology experience, which is highly significant given the increased reliance on e-learning. The overarching principle is that older learners vary from younger learners in their needs and preferences, and for them to successfully engage in meaningful learning these needs and preferences must be accounted for in the design of training and instructional programs. At the same time, it is important to recognize that for the most part improving the usability of training and instructional programs for older adults will likely result in improvements of these programs for people of most ages.

1.2   Societal trends

1.2.1   The technology explosion

Coupled with the aging of the population, there are a number of important societal trends that highlight the importance of well-designed training and instructional programs for older adults. One highly significant trend is the technology explosion. Use of technology is pervasive and has become an integral component of work, education, communication, routine services, and entertainment. For example, most workers, including those in non-service sector occupations, use some form of technology in the daily performance of their jobs. Technology is reshaping production processes and the task content of jobs, as well as changing the skill requirements of jobs. Workplace technologies are also evolving and workers need to upgrade their skills and knowledge continually to remain competitive in the workplace. Use of automatic teller machines, interactive telephone-based menu systems, service kiosks, intelligent transportation systems, and "smart" mobile devices is also quite common. In many residences, common home tasks such as using security and entertainment systems and controlling temperature also require learning to use some type of technology.

Technology devices are also becoming more integrated and providing faster and more powerful interactive services. Large and increasing numbers of people have almost immediate direct access to a wide array of information sources and services. Internet use in the United States has become pervasive: 78% of adults age 18+ use the Internet (Figure 1.2) including 42% of those aged 65 and older. In addition, 85% of Americans aged 18 and over own a cell phone including 68% of older adults (Pew Internet & American Life Report, 2011a), and half of adult cell phone owners have apps on their phones for a wide variety of information and functions (Purcell, 2011). Communication via e-mail and social network sites is also quite prevalent. In fact, one of the primary reasons older adults use the Internet is to use e-mail. E-books and e-readers such as Kindles or Nooks are also becoming popular.

Technology is also being increasingly used within the healthcare arena for service delivery, in-home monitoring, interactive communication (e.g., between patient and physician), transfer of health information, and peer support. For example, there are myriad health websites available that provide consumers with access to health information and services and the ability to buy medical supplies, equipment, and even medications or supplements. To use these health websites consumers need to learn numerous skills such as basic mouse and window skills, how to search for relevant websites, filter search results, and navigate within websites.

Finally, technology is influencing the implementation and format of training and instructional programs. *E-learning*, which refers to online or computer-based instruction (Chapter 10), is quickly emerging as a

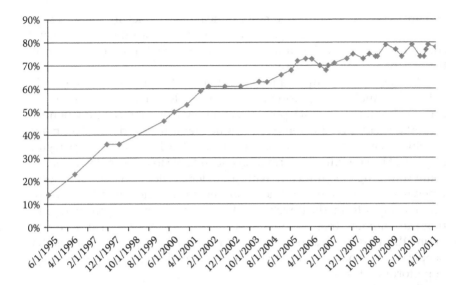

Figure 1.2 Percentage of adults in the United States online, 1995–2011. (From Pew Internet & American Life Project, 2011b, on Internet adoption.)

preferred training method for a variety of situations, such as worker training or learning to use new technologies such as mobile phone devices. Online courses sponsored by organizations such as AARP, SeniorNet, or universities are also quite common. To date there has only been limited research examining whether mature learners can learn effectively via e-learning formats. Also, most e-learning programs involve multimedia formats (audio, text, video, and animation) that may prove challenging for older learners (Chapter 10). This is another fruitful area for investigation.

Overall, technology is used in practically every aspect of everyday living. This technology explosion mandates that to function independently and successfully engage in routine activities, older adults need to interact with some form of technology on a routine basis. This implies that they will need to engage continually in learning and training to operate and maintain technology systems and to keep abreast with existing and emerging technology trends.

1.2.2 Trends in healthcare and the workplace

There are also changes occurring within healthcare that result in a tremendous need for training and instruction. For a variety of reasons, healthcare is increasingly occurring in settings such as the home rather than in professional medical settings. Patients and their families are being asked to assume an increasing role in the management of their own health and are expected to perform a range of healthcare tasks and interact with a vast array of medical devices and technologies within their home and community settings. This implies a great deal of learning on the part of the patient and family caregivers.

As noted, part of this change in healthcare delivery is a focus on technology. Technology applications are commonly used within home settings to monitor a patient's physical, emotional, or cognitive functioning. With the rapid introduction of electronic medical records (EMRs), many of which have patient portals, consumers also have access to varying degrees of their medical information and are able to perform tasks such as communicating electronically with providers, scheduling appointments, renewing prescriptions, and accessing health management information through links to medical websites. Furthermore, there is an increasing trend to mandate that consumers enroll in benefits and insurance programs through the Internet. The use of all of these systems implies that people not only have to learn to use these technologies, but also new ways of performing tasks.

These changes in healthcare have broad implications for older adults as they are more likely to need and engage in healthcare activities. Generally, the prevalence of chronic conditions or illnesses such as dementia, diabetes, heart disease, or stroke increases with age, and consequently older adults (especially the oldest old) are more likely to need some form of care or assistance. Family members, many of whom are also older adults, are often the primary and preferred source of help for elders with an illness or chronic condition. It is not uncommon for these older caregivers to have a chronic condition also. Having an illness or disability or even the stress from caregiving can have an impact on one's capacity to learn. Furthermore, many older people live alone and have limited access to support. These issues present challenges with respect to developing effective training strategies for use and maintenance of medical devices and technologies and performance of healthcare tasks.

There are also enormous changes in work environments and organizations and these changes, which are expected to continue in the future, have created new knowledge, skill, and ability requirements for workers. For example, ongoing developments in technology are reshaping production processes and the task content of jobs, as well as changing the skill requirements of jobs. In the future, the rapid pace of technological change and the transition to a knowledge-based economy is going to increase the demand for highly skilled and well-educated workers. We can also anticipate that developments in technology will continue to shape what is produced, how material and labor input are combined to produce it, how work is organized, and the content of work. Again, this means that workers will continually need to upgrade their skills and knowledge to remain competitive in the workplace. Issues of skill obsolescence and training are especially significant for older workers, as they are less likely to have had exposure to technologies such as the Internet, and data suggest that employers invest less in training older workers than younger workers (Czaja and Sharit, 2009).

Other changes in work environments that have important implications with respect to the need for training include changes in organizational structures that now have a focus on decentralized decision making and collaborative work. In these types of structures workers are often confronted with a need to learn entire processes as opposed to specific jobs, and to be able to communicate effectively with diverse teams of people, usually through the use of technology, who may be in distant locations. The prevalence of home-based work and telecommuting is also increasing. Overall, these organizational and job changes generally require that workers need to receive retraining to keep current with job requirements. These changes also raise important issues with respect to how workers, especially the increasing number of older workers, will receive the training and education needed to update their skills.

1.3   Content of the book and approach to the topic

1.3.1   Overview of the content

As noted earlier in this chapter, the goals of this book are to present a state-of-the-science summary of the topic of training and instructional design for older adults and, where possible, to present some basic principles and guidelines regarding best practices for designing training and instructional programs for older people. It should be pointed out that although there is a substantial literature that focuses on adult cognition and adult learning, there are no distinct theories of instructional design for older people. Instead, there are theories, models, and perspectives concerning how people learn or acquire knowledge and skills, and how instruction should be formulated and delivered. There are also some general guidelines and recommendations for training and instructional programs that have been derived from these theories and from empirical investigations. However, there is no set formula for how training and instructional programs should be designed and delivered for all older adults for all learning situations. Our approach is to summarize current thinking regarding training and instructional design and demonstrate, based on existing knowledge about aging, how these findings can be applied to guide the development of training programs for older people. Of course, we also base our recommendations on the rather limited amount of research that is available on training and older adults.

Given the complexity of the topic, this book covers a wide range of issues. Consistent with the general principles of human factors, we believe that good design is predicated on understanding the characteristics of the target population. Thus, in Chapter 2 we discuss the demographics and the characteristics of older learners and the relevance of these characteristics to the design of training and instructional programs. Chapter 3 provides an overview of what is generally known about training and older adults, which is followed by a general discussion of learning and skill acquisition and includes information on factors that affect the learning and skill acquisition processes (Chapter 4). Important to the concept of initial learning are the topics of retention of what has been learned outside the training situation and the ability to transfer the learned information to a variety of situations (Chapter 5). Chapter 6 focuses on how factors such as a learner's motivation, anxiety, and fatigue have an impact on initial learning of new material and retention and transfer of that material outside the training situation. A learner's motivation for engaging in training is influenced by a variety of factors and is critically important to learning success. Anxiety about ability to learn new skills and concepts, as well as fatigue, also have a pronounced impact on the learning process as they influence capacity for learning. These issues are particularly important for older adult learners.

The next section of the book begins a discussion of approaches to instructional design. Chapter 7 is a lead-in to this section and presents an introduction to the human information-processing system to provide a framework for the material that follows. This chapter also includes the presentation and discussion of a human information-processing model that is more directly relevant to issues older learners are likely to face. Using this framework, Chapter 8 discusses various methods and approaches to instructional design, with special emphasis given to the need for sequencing instructional material properly and the four-component instructional design (4C/ID) model. This is followed by a chapter that reviews more systems-based approaches to designing instructional programs (Chapter 9). We then examine issues with respect to new training and learning formats such as multimedia and e-learning programs (Chapter 10), followed by a discussion of issues critical to the evaluation programs and methods that can be used for program evaluation (Chapter 11). In Chapter 12 we provide examples within domains such as work and healthcare where training is and will continue to be a critical issue for older adults. We also provide a synthesis of the main themes that emerged throughout the book.

1.3.2   Overview of the approach

Given the enormity of the information on the topics addressed in this book and the vast and continually emerging research in this area, we cannot provide comprehensive coverage of the broad range of issues that are discussed. Instead, we attempt to summarize what is known and how this applies to the development of training programs for older adults. In some cases, such as e-learning and multimedia formats, we also point out where there is a need for more information. Clearly, this book does not provide a prescription for how to design training and instructional programs for all older adults in all learning situations. Rather, our hope is that we have illustrated the importance of this topic for current and future generations of older people and highlighted some of the issues that need to be considered by designers of training programs and researchers in this area.

Finally, the book is intended for a broad audience from designers of training and instructional programs and managers of older workers, to undergraduate and graduate students who have interests in aging, human factors, cognition, and training. Our approach is to make the information accessible to people with varying backgrounds and not be overly technical. Thus the book is not written as a typical academic text with extensive references to the literature. Our intention was neither to present a high-level research exposition on the topic of training and instruction for older adults, nor to provide an overly simplified presentation of this topic. Rather, our goal was to offer something more middle-ground in nature that refers to the literature, and in some cases may even overview some studies. For those who want more depth regarding the various topics we provide recommendations for further reading. Also, for more explicit guidelines concerning designs of printed words, icons, labels, and similar artifacts for older adults, a number of human factors texts are available (e.g., Fisk et al., 2009; Pak and McLaughlin, 2010) that provide useful information. Similarly, guidelines for designing websites so that older adults can more easily interface with them also address many fundamental human factors issues and can be found in a number of sources (e.g., The National Institute on Aging [http://www.nia.nih.gov/health/publication/making-your-website-senior-friendly] and the National Library of Medicine [http://www.nlm.nih.gov/pubs/checklist.pdf]).

Recommended reading

Swezey, R.W. and Llaneras, R.E. (1997). Models in training and instruction. In G. Salvendy (Ed.), *Handbook of Human Factors and Ergonomics*, 2nd ed. New York: John Wiley, 514–577.

Willis, S. (2004). Technology and learning in current and future older cohorts. In R.W. Pew and S.B. Van Hemel (Eds.), *Technology for Adaptive Aging*. Board on Behavioral, Cognitive, and Sensory Sciences, Division of Behavioral and Social Sciences and Education. Washington, DC: National Academies Press, 209–229.

chapter two

Characteristics of older adult learners

2.1 Overview

The objectives of this book are focused on understanding issues related to learning and training and ultimately to improved design of training and instructional programs for older adults. Our approach to these issues is based on a human factors approach. The discipline of human factors is concerned with understanding interactions among humans and other elements of a system and applies theory, principles, data, and other methods to design in order to optimize human well-being and overall system performance (Human Factors and Ergonomics Society, 2012). A central tenet of human factors is that the characteristics of user populations must be considered in the design of products, tasks, environments, and programs that people use. Consistent with the human factors approach, in this chapter we provide an overview of some empirical evidence about the characteristics of older adults.

We start with a few important caveats. One is that the definition of *older adults* varies depending on the context of the discussion. For example, within the workplace people aged 55 and older are typically considered *older workers*, whereas in much of the scientific literature those aged 65 and older are considered older adults. Today, given the vast numbers of people who are living into their 80s and 90s and beyond, an important distinction is that between younger older adults, those aged 65–80 years, and those aged 80+. Someone in their 60s or 70s is typically very different from someone in their 80s. There are also important differences in characteristics between those who are 80–89 and those who are 90+ years. These distinctions are becoming extremely important as the number of people living age 90 and above is increasing. Another important caveat is that aging is associated with tremendous heterogeneity and all older people are not alike due to differences in genetics and experiences over the life course. A third important caveat is that aging is associated with plasticity. Older adults can experience improvements and gains in physical, cognitive, and functional performance and can learn new skills. They also bring a wealth of knowledge, skills, and experiences to situations.

2.2   Demographic profile

2.2.1   Age, gender, and ethnicity

As noted in Chapter 1, not only is the percentage of older adults in the population increasing but also the older population itself is getting older. The current population of people aged 90 and older is about two million and is expected to quadruple over the next few decades. By 2050 people aged 65 and older will represent about 20% of the population and those aged 90+ will represent 2%. These trends are being paralleled in other countries throughout the developed world. The increase in the oldest old will continue to have a significant societal impact on healthcare, retirement and pension systems, and living and family arrangements. People in the later decades are more likely to have disabilities such as arthritis, and more significant vision and hearing impairments. Memory impairments also become more likely with increased age. They are also more likely to have mobility restrictions and live alone with limited informal support. As noted throughout this chapter, these changes have implications for the structure, format, and delivery of training and instructional programs and support materials. For example, structural barriers related to outreach, scheduling, and transportation often impede the ability of older adults to participate in community educational programs. In Chapter 10, we discuss the topic of online learning programs as a potential solution to these issues.

In terms of gender, older women outnumber older men, especially in the later decades. In 2009, there were 22.7 million older women and 16.8 million older men, or a ratio of 135 women for every 100 men. The female-to-male ratio increases with age. For example, for the population aged 90 and over, the number of men per 100 women is about 35 (U.S. Census Bureau, 2011). Most women in this older age group are widowed.

Consistent with demographic changes in the U.S. population as a whole, the older population is becoming more ethnically diverse. The greatest growth will be seen among Hispanic persons, followed by non-Hispanic blacks. Currently, individuals from ethnic minority groups are less likely to own or use technologies such as computers. This has vast implications for the design of training and instructional programs. One obvious implication is language. For example, currently many older Hispanic adults have limited fluency in English; thus, to ensure that training and instructional programs are accessible to this population they need to be available in Spanish. The same is true for other ethnic groups. In addition, it is important for instructors to have some background in the cultural/ethnic mores of minority populations. Also, although use of the Internet is increasing among older adults, older minority adults are less likely to have access to the Internet and technology skills. This implies that training and instructional programs that teach technology skills need to be targeted for older minority populations. Finally, literacy is lower among older adults in general but especially minority older adults. Therefore, it is important that language used within course instruction and in training manuals is not overly complex and highly technical as this would create problems with comprehension and may also cause frustration and anxiety and will ultimately limit the success of training programs.

2.2.2   Health status

On some indices, today's older adults are healthier than previous generations. The number of people 65+ reporting very good health and experiencing good physical functioning, such as the ability to walk a mile or climb stairs has increased in recent years. Disability rates among older people are also declining (Federal Interagency Forum on Aging-Related Statistics, 2010). However, as noted above, the likelihood of developing a disability or chronic condition increases with age, especially among those in the later decades (Figure 2.1). About half of people over the age of 65 report problems performing basic activities of daily living and this number increases in the oldest old. For example, the proportion of

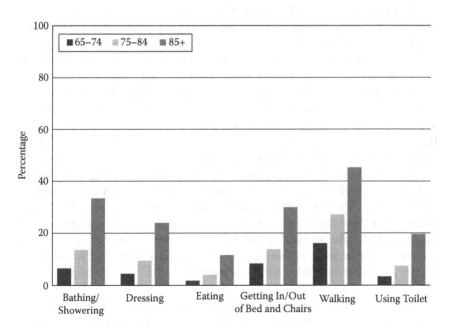

Figure 2.1 Percentage of persons with disability in activities of daily living by age group: 2007. (From A Profile of Older Americans: 2010. Administration on Aging. U.S. Department of Health and Human Services.)

people aged 90–94 having disabilities is 13% higher than those aged 85–89 (U.S. Census Bureau, 2011). Many older adults also have chronic conditions such as hypertension, diabetes, heart disease, arthritis, or dementia. These conditions contribute to declines in quality of life and functioning. As expected, the prevalence of these conditions is higher in the older cohorts. In addition, prevalence of many chronic conditions varies widely by race and ethnicity. Non-Hispanic white older adults are more likely to report good to excellent health and have lower rates of chronic conditions than non-Hispanic blacks and Hispanic older adults.

2.2.3   Living arrangements, education, occupation, and leisure activities

In terms of living arrangements, contrary to popular beliefs, the majority of older adults live independently in the community. In fact, only 1% of those aged 65–69 and 3% of those aged 75–79 live in nursing homes. Although this number increases with age, it is still fairly low among those aged 85–89 (11%) and 90–94 (20%; U.S. Census Bureau, 2011). Many older adults, especially older women live alone and others live with family members or in some type of group quarters.

Today's cohort of older adults has higher educational achievements than previous generations and this will continue with the aging of the baby boomers. In 1965 only 5% of those 65 and older had a bachelor's degree, and by 2008 this number had increased to 21%. Interestingly, in 2008 about 14% of those 90 and older had a college degree (Federal Interagency Forum on Aging-Related Statistics, 2010; U.S. Census Bureau, 2011). Educational attainment is related to a person's well-being and health and the ability to engage in activities such as lifelong learning. In the upcoming decades the number of older people with advanced degrees will increase, which has tremendous implications for the development of training programs as these older adults are likely to continue to engage in learning activities. In fact, recent data from a survey conducted by the American Association of Retired Persons (AARP, 2000) indicated that the majority of older people included in their sample were interested in continuing education to keep up with worldwide developments, personal growth and fulfillment, and the simple joy of learning something new. Opportunities for social interaction also motivate older adults to participate in learning and training activities.

With respect to work and employment, many adults in their middle and older years are choosing to remain in the workforce longer or return to work because of concerns about retirement income, healthcare benefits, or a desire to remain productive and socially engaged. Current labor projects indicate that by 2025 the number of workers aged 55+ will be about

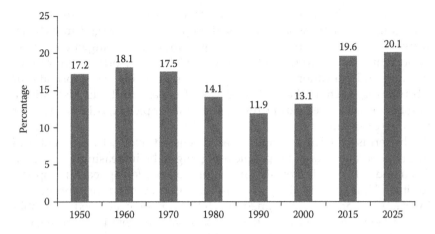

Figure 2.2 Percentage of the labor force aged 55+, 1950–2025. (From U.S. General Accounting Office, 2001.)

33 million (U.S. General Accounting Office, 2003). There will also be an increase in the number of workers aged 65+ (Figure 2.2). Trends in retirement patterns are also changing. Rather than take up full-time leisure, workers and retirees in their 50s, 60s, and 70s are increasingly seeking more options such as special projects or contract work, the opportunity to start a second (or third) career, or to start a new business. All of these trends underscore the need for training older adults.

To ensure that older adults are able to compete successfully in today's labor market and adapt to new workplace demands and technologies it is important that older adults are provided with access to retraining programs and incentives to invest in learning new skills. In today's work environments, training and retraining of older workers is a critical component of organizational effectiveness. The importance of providing training for older workers is due to a number of changes in the workplace, and in particular, ongoing developments in technology, which are reshaping work processes and the task content of jobs, and changing the skill requirements of jobs. We are also transitioning to a knowledge-based economy where there is an increasing demand for highly skilled and well-educated workers. Also, in most organizations there is an increased focus on decentralized decision making and collaborative work, which implies that many workers are confronted with a need to learn entire work processes as opposed to specific jobs. They must also be able to communicate effectively with diverse teams of people. Older workers recognize these changes and understand that having the opportunity to refresh their current skills or learn new skills is an important aspect of their job and continued employment opportunities.

Coupled with these trends is the fact that work arrangements are changing and many workers are self-employed or engaged in contract work or telecommuting. These changes in work arrangements raise important issues with respect to how nonstandard workers will receive training and education needed to update their skills. New formats for job training, such as the increased use of e-learning tools, also raise a number of issues regarding the design and implementation of worker training programs.

In terms of leisure activities, as discussed above, the current and future generations of older people are going to be increasingly well educated and active and interested in the pursuit of lifelong learning opportunities. Thus there is likely to be an increased demand for training and instructional programs designed to meet the preferences, needs, and abilities of older people. In the following sections we provide a summary of normative age-related changes in abilities that have relevance to training and instruction, and in the next chapter we discuss the implications of these changes for the design of training programs.

2.2.4 Use of technology among older adults

Given the ubiquitous use of technology in most training and learning environments, we present some overview information on older adults and the use of technology as it is important when thinking about the structure or format of training and instructional programs. In this regard, although use of technology such as computers and the Internet among older people is increasing, it is still lower than that of younger age groups. In 2010, about 42% of people age 65+ were Internet users as compared to 78% of people age 50–64 and 87% of those 30–49 years old. Among those 65 years and older only about 25% of those 75–84 years of age and 5% of those 85+ years are computer or Internet users (Charness, Fox, and Mitchum, 2011). Furthermore, people aged 65+ are much less likely than younger people to have a high-speed Internet connection. In 2010, 31% of adults aged 65+ in the United States had broadband access at home as compared to 75% of those aged 30–49 years and 61% of those aged 50–64 years. In addition, seniors who do have high-speed Internet access tend to be white, highly educated, and living in households with higher incomes (Pew Internet & American Life Project, 2004, 2011b). Recent data from our group also indicates that older adults are less likely than younger adults to use other forms of technology such as DVDs. Use of technology also tends to be lower among people with chronic conditions. As discussed above, the likelihood of having one or more chronic condition increases with age.

However, our data and those of others strongly indicate that older adults are willing and able to use technology systems. Barriers to use include lack of access and knowledge about potential benefits, lack of technical support, cost, fear of failure, and complex interfaces or interfaces that are designed without considering the needs of older adults (e.g., small fonts). With respect to the design of training programs, these findings indicate that many older adults are still in need of basic technology training, and that technology systems used in training programs need to be designed to accommodate the needs and preferences of older people.

2.3   Abilities and older adults

2.3.1   Sensory/perceptual systems

Sensory/perceptual skills are critically important to training and learning. As discussed in our chapter on human information processing (Chapter 7), when people are engaged in a learning/training task they receive information from the environment and must perceive, comprehend, and act on that information. In most learning situations a person is presented with information from an instructor or peer, training manual, or display screen. This information is generally presented in an oral, visual, tactual (e.g., Braille), or multimodal format. The person receives this information from a sensory modality such as vision, hearing, or touch, or some combination of these modalities.

For example, imagine a training class where workers are learning to use a new software program and they are listening to an instructor guide them through practice examples that are presented in a training manual and on computers, or a person at home learning a new language through a multimedia training package that includes auditory and a variety of types of visual information (e.g., text, animation). In both situations the learners rely on the visual and auditory sensory modalities to receive the information. They also rely to a lesser extent on the tactual sense to receive information such as feedback regarding use of the keyboard or mouse, or perhaps to a greater extent if they are blind or extremely visually impaired and information is presented in Braille. Unless attention is given to the characteristics of older adults and age-related changes in sensory/perceptual abilities (Figure 2.3), these situations can be challenging for older people and affect their ability to learn.

2.3.1.1   Vision

There are a number of changes in visual abilities that occur with aging that are relevant to the design of training and instructional programs. Currently, about 14 million people in the United States suffer from some type of visual impairment and the incidence of visual impairment increases with age. Generally, with increased age there is a decline in visual acuity, which is the ability to resolve detail. Older adults also experience declines in accommodation, which is the ability of the eyes

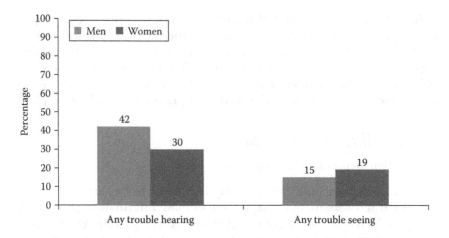

Figure 2.3 Limitations in hearing and seeing among population 65+, 2008. (From Federal Interagency Forum on Aging-Related Statistics, 2010.)

to adjust to different viewing distances. It takes older people longer for their eyes to adjust to shifts in viewing distances such as information presented on a whiteboard and information presented on a computer screen. With age, there is also a loss in contrast sensitivity (the ability to distinguish a target from a background based on differences in color brightness), decreases in the ability to adapt to differences in levels of light or darkness, declines in the ability to distinguish colors (especially in the blue region), and heightened susceptibility to problems with glare. However, older people need more light than do younger people to perceive the same information; thus the level, location, and type of lighting is very important.

Visual search skills and the ability to detect targets against a background also decline as people grow older. Many of these changes begin in the 40s and become more pronounced in the later decades. Although the majority of older people will not experience severe visual impairments, they may experience declines in eyesight sufficient to make it more difficult to perceive and comprehend visual information. Even with correction it is generally more difficult for older adults to read small text or perceive small objects, to discriminate subtle differences in brightness or color, and to adapt to changes in viewing distances and levels of light. Also, the incidence of visual impairments such as cataracts and macular degeneration also increases with age. Clearly, age-related changes in vision have implications for the presentation of visual information such as information in instruction manuals or presented on display screens, and for the design of training environments (Chapter 3).

2.3.1.2 Audition

Aging is also associated with changes in hearing ability. Many older adults experience some decline in auditory functioning. Recent data indicate that over 40% of older men and about 30% of older women report trouble hearing. These percentages are higher for the older age groups (The Federal Interagency Forum on Aging-Related Statistics, 2010; Figure 2.3). Specifically, age-associated losses in hearing include: a loss of sensitivity for pure tones, especially for high-frequency tones; difficulty understanding speech, especially if the speech is distorted; problems localizing sounds; problems in binaural listening; and increased sensitivity to loudness. It also takes older people longer to process auditory information. For example, it is difficult for older adults to hear high-pitched sounds or understand speech that is rapid or distorted, especially in noisy environments. Older people may also find it difficult to understand synthetic speech, as this type of speech is typically characterized by some degree of distortion. These changes in audition are also relevant to the design of training programs. For example, multimedia systems that include a speech component may be problematic for older adults if the rate of speech is too rapid or is distorted (Chapter 3).

2.3.1.3 Motor skills

Older adults also experience changes in motor skills. These changes include slower response times, declines in ability to maintain continuous movements, disruptions in coordination, loss of flexibility, and greater variability in movement. Generally, older adults are 1.5 to 2 times slower than younger adults. They are also prone to more movement errors. The incidence of chronic conditions such as arthritis that affect movement also increases with age (Figure 2.4). These changes in motor skills have direct relevance to training and instructional design, especially given the increased use of technology in instructional settings. For example, various aspects of mouse control such as moving, double-clicking, fine positioning, and dragging are likely to be difficult for older people. In fact, we have found that mastering the use of a mouse can be more challenging for older people than learning to use a software program. Therefore, it is important to ensure that older adults who have limited computer experience master basic mouse and window functions before proceeding to other computer-based learning activities. Alternative input devices such as a touch panel interface might also be beneficial for older people.

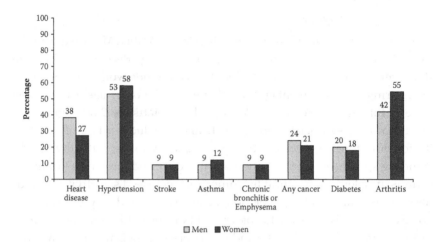

Figure 2.4 Chronic health conditions among the population age 65+, 2007–2008. (From Federal Interagency Forum on Aging-Related Statistics, 2010.)

2.3.2   Cognition

2.3.2.1   Memory

There are numerous age-related changes in cognitive abilities that have relevance to learning and the design of training programs. This section provides a summary of these changes and they are also discussed in other chapters throughout the book (e.g., Chapters 3 and 7). At the outset it is important to re-emphasize that age-related changes in cognition do not mean that older people cannot learn. Also, although the likelihood of developing a memory impairment such as Alzheimer's disease or another form of dementia increases after age 65, not all older people develop or suffer from dementia. As noted earlier in this chapter, aging is associated with a great deal of plasticity. However, it is important to have some understanding of the nature of age-related changes in cognitive abilities so these changes can be accommodated in training and instructional programs.

Generally, there are a number of aspects of cognition that decline with age. As discussed in Chapter 7, a typical distinction when discussing aging and cognition is between crystallized abilities, which reflect knowledge such as language skills and knowledge about a particular job or topical area, and fluid abilities, which are involved in active processing of current or new information. Fluid abilities are very important to new learning and skill acquisition. It is well established that crystallized abilities tend to remain stable until the very later decades, whereas fluid abilities tend to decline with age (Figures 2.5 and 2.6).

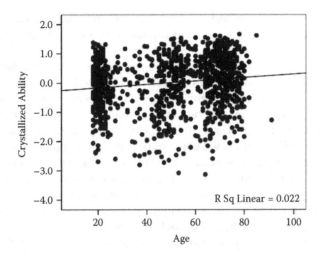

Figure 2.5 Age-related changes in crystallized cognitive abilities (*N* = 1,197). (From Data from the project described by Czaja et al., 2006.)

In terms of specific abilities, it is generally thought that memory declines with age. However, this is an oversimplification. Memory is complex and there are different types of memory. Working memory, which refers to the ability to temporarily hold information while we process it or use it, declines with age. For example, imagine that you are listening to someone giving you instructions on how to use a new version of Microsoft Word while you are editing a document. In this case you are

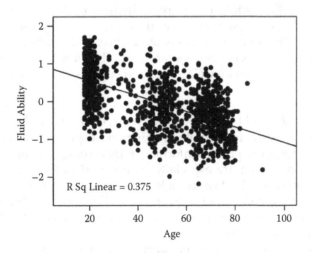

Figure 2.6 Age-related changes in fluid cognitive abilities. (From Data from the project described by Czaja et al., 2006.)

relying on working memory to hold the instructional information and match it to your editing task. Working memory plays an important role in learning.

Another type of memory is long-term memory, which refers to the more permanent memory storage system. This is the memory store for knowledge. There are different types of long-term memory. Semantic long-term memory refers to the storage system for factual information that has been acquired over a lifetime. Generally, this aspect of memory shows little decline with age. People may experience some problems with retrieval of semantic information if the information has not been used or activated in a while, but generally the information is not lost and can be retrieved if people are provided with the appropriate cues. In terms of training, it is good to try to build on or make use of someone's semantic knowledge such as familiar constructs or concepts.

Prospective memory or remembering to do something in the future is another type of long-term memory. We can distinguish between two types of prospective memory. Time-based prospective memory is remembering to do something at a later time such as show up for a training class tomorrow at 2 p.m. Event-based prospective memory is remembering to do something after some event such as meeting a tutor after work. There are age-related declines in both types of prospective memory, but declines are typically greater for time-based prospective memory.

Finally, there is procedural long-term memory or remembering how to perform some activity such as driving or using a software program or some work procedure. The degree of age-related declines in procedural memory depends on how well learned the procedure is (i.e., the level of automaticity). Older adults do not typically have difficulty remembering procedures that are automatic or well-learned and they can learn new procedures with sufficient amounts of practice. It also takes older adults longer to unlearn procedures. As far as possible, it is best to make new procedures consistent with well-learned procedures to minimize the need for unlearning.

2.3.2.2   Rate of information processing, attention, and reasoning
One well-established finding about aging and cognition is that it takes older people longer to process information than younger people. Age differences in processing are especially pronounced in the older decades. They also increase with task complexity. As expected, it may also take older adults longer to respond. In training situations it is important to allow older adults sufficient time to process learning material and to make responses (Chapters 3 and 4).

Attention, which refers to the capacity for processing information, is another important aspect of cognition, especially for learning and training. We know that humans have limitations with respect to the amount of information that they can process at any one time, and when capacity is exceeded or overloaded performance tends to decline and be error-prone. Also, performing tasks at near-capacity levels for extended periods of time can be fatiguing, especially for older adults. There are different aspects of attention. Selective attention refers to the ability to filter information and focus on selected information in the presence of other information. For example, someone can select and focus on relevant pieces of text when searching through a webpage even though there might be advertisements on the page. This becomes harder to do with increased amounts of clutter and irrelevant information, especially for older adults. It is also easier to select relevant information if it is salient and stands out in some way. For example, it is easier to find information within a page of an instruction manual if it is organized into sections with boldface subheadings. Older adults are especially susceptible to problems with clutter and extraneous information.

Divided attention is the ability to divide our attention between competing activities or multiple sources of information, for example, listening to an instructor while following along on a computer screen or reading an instruction manual while listening to background music. A classic but dangerous example is driving while talking on a cell phone. Generally, the ability to divide attention decreases with age, especially for complex tasks.

There are also age-related differences in reasoning abilities or the ability to draw inferences from newly presented information. Older adults tend to do less well on these types of tasks than do younger adults. However, age-related differences are reduced if familiar concepts are used. For example, we found that older adults were more successful learning to use a new text-editing program when we used analogies to existing concepts such as file drawers and paper file folders. We have also found that reasoning ability is important to the performance of Internet-based information search tasks and the ability of individuals to integrate information across a variety of information sources. In most training and learning situations people need to integrate information from different sources such as an instructor and an instruction manual. They must also integrate new information with existing knowledge. This can be challenging for older people if there are too many sources of information, the information is complex and highly unfamiliar, or they have insufficient time. We have also found that providing older adults with strategies to help them organize or integrate information is beneficial.

Of course there are other factors that affect the ability to learn new skills and that affect participation in training and instructional programs, such as motivation, anxiety, or discomfort about being in new learning situations, and self-efficacy, which refers to the belief that one is capable of learning something new. These factors are discussed in Chapter 6.

2.4 Summary

As shown throughout this chapter, older adults represent a very heterogeneous group, varying in demographic characteristics such as ethnicity, culture, educational background and economic circumstances, technology experiences, and occupational backgrounds. They also vary in health status and skills and abilities. In addition, there are significant differences in abilities according to cohorts within the older adult population. It is important to note that aging is also associated with considerable plasticity and the functional abilities of older adults can experience growth and older adults can learn. Designers of training programs need to understand the heterogeneity of the older adult population and ensure that training and instructional programs are designed to accommodate this diverse group of users, and that pilot testing and evaluation of training programs are conducted with representative older adult user groups. A summary of the information discussed in this chapter is presented below.

- Adults in their 60s and 70s are typically vastly different from those in their 80s, and there are important differences in characteristics between those who are 80–89 and those 90+ years.
- Aging is associated with tremendous heterogeneity and all older people are not alike.
- Older adults can experience improvements and gains in physical, cognitive, and functional performance and can learn new skills.
- Older adults bring a wealth of knowledge, skills, and experiences to situations.
- The likelihood of having a chronic condition, disability, or mobility impairment increases with age.
- Many older adults, especially older women, live alone and have limited access to support.
- Literacy tends to be lower among older adult populations, especially ethnic minorities or those from lower socioeconomic strata.
- Today's cohort of older adults has less technology experience and uses technology less than younger adults.
- Older adults are interested in maintaining active and productive lives and engaging in new learning activities. Many older adults need or want to be engaged in some form of employment.
- There are normative age-related declines in vision and audition.
- There are age-related changes in motor skills and speed of responding.
- Cognitive abilities such as working memory and attention decline with age, whereas others such as knowledge show little age-related decline.
- Rate of processing information declines with age.
- Age-related changes in cognition are greater for complex tasks or for unfamiliar information.

Recommended reading

Birren, J.E. and Schaie, K.W. (2005). *Handbook of the Psychology of Aging*, 6th ed. New York: Academic Press.

Fisk, A.D. and Rogers, W.A. (1997). *Handbook of Human Factors and Older Adults*. Orlando, FL: Academic Press.

Hofer, S.M. and Alwin, D.F. (2008). *Handbook of Cognitive Aging: Interdisciplinary Prospective*. Newbury Park, CA: Sage.

Park, D. and Schwartz, N. (2008). *Cognitive Aging: A Primer,* 2nd ed. Philadelphia: Psychological Press.

chapter three

Training older adults
An overview

3.1 Introduction

As noted in Chapter 1, older adults continually need to engage in learning in order to use new products and devices and perform new tasks and activities at work, home, and in service environments. Many older people also engage in learning activities to acquire new hobbies, to remain intellectually challenged, or to brush up on a previously learned skill. For example, AARP (2000) conducted a survey of over 1,000 adults aged 50 and older and found that the majority of people were interested in life-long learning so that they could advance their skills, experience personal growth, and for the simple joy of learning something new. They also expressed interest in learning about a variety of topics such as hobbies, health, and nutrition. There are also numerous online programs geared for seniors, and organizations such as AARP offer a large number of educational programs for older adults. In all of these cases engaging in new learning typically involves some form of training. As the terms *learning* and *training* are often used interchangeably, for clarity we refer to learning as the acquisition of new or existing knowledge, skills, behaviors, or preferences, whereas training is the process or instructional activity aimed at imparting knowledge, skills, behaviors, or preferences.

As discussed throughout this book, the structure and style of training programs can vary on a number of dimensions. For example, training may take a variety of forms including individual training sessions, group training sessions, workshops, online instruction, an instruction manual, or informal training from a coworker, family member, friend, or some combination of these. Training programs also vary in duration, frequency, length of training sessions, structure (e.g., passive, interactive), and other factors such as the protocols used for evaluation. Training can also be self-paced or paced by an instructor or a software program, and can occur in a variety of settings such as work, at home, in a classroom (e.g., community college), community center, or some distance-learning location. The results from the AARP survey indicate that preferences of older adults with respect to training format vary according to the topic being studied. However, most respondents indicated that they preferred small group settings or learning on their own. In addition, the respondents indicated they preferred *hands-on* active learning approaches and learning methods that are easy to access and did not involve large investments of time. Data from our research also indicate that the preferences of older adults with respect to training format vary, but most prefer individualized or small group instruction with peers.

Irrespective of the topic or the format, the general goal of any training program is to ensure that the learners or trainees can use the instructional material successfully at a later point in time or transfer what they learned to a new situation. Therefore, it is critical to ensure that the learning that occurs in the training environment does not just reflect immediate mastery but also transfers to behavior in other settings outside of training (Chapter 5). Generally this requires that training programs must promote the coordination of all skills required for a task, the integration of new skills with previously acquired knowledge and skills, and the ability of the learner to apply the new knowledge and skills across a variety of situations. For example, when teaching someone basic windowing and mouse skills, it is important to teach all of the essential concepts such as sidebars, search boxes, scrolling, links, and how to use the mouse in relation to these items, for example, right- and left-clicking, double-clicking, and cursor positioning. It is also important to impart to the learner how to rely on previous knowledge about a topic to find search terms, and to use common analogies to impart new terminology. Finally, the learner must understand when these constructs or skills are applicable and be able to use them across many different window applications or websites. It is important to note that they must also learn to discriminate when skills or constructs do not apply, such as when double-clicking is not necessary.

Much has been written on how to design optimal training and educational programs for various learner populations (e.g., children, adults, athletes, medical students) and for various types of skills and activities (e.g., new language, sporting activity, software application). There also has been research directly aimed at identifying training techniques that are optimal for older learners. Review of this extensive literature is well beyond the scope of this book. We do, however, provide a general discussion of training methods and approaches in Chapters 8 and 9. Also, given the increased emphasis on the use of technology as an instructional medium, we include a separate chapter on e-learning and multimedia training formats (Chapter 10). Our goal in this chapter is to provide some basic principles and guidelines, based on the current literature, regarding best practices for training older adults. To set the stage, we begin our discussion by reviewing some basic facts about learning and aging. A more detailed review of aging and skill acquisition is presented in the following chapter.

3.2   Aging and learning: An overview

3.2.1   Individual differences

One important construct when discussing any aspect of the aging process is the source of performance variability. Generally, studies examining aging and cognition or aging and learning distinguish between the following two sources of performance variability: interindividual variability in performance and intraindividual variability in performance. One type of interindividual variability is that which is observed in between-group comparisons such as cross-sectional studies where differences between various age groups are being compared at a given measurement point. For example, Figure 3.1 depicts interindividual variability across different age groups for a variety of cognitive measures. In cross-sectional studies, data are typically pooled across individuals within a group. Sometimes differences among age groups that are found in these studies are not entirely associated with the aging process but reflect cohort or generational differences. In many of our studies comparing performance of younger and older adults on computer tasks we found that some of the observed variability in performance between the groups was accounted for by differences

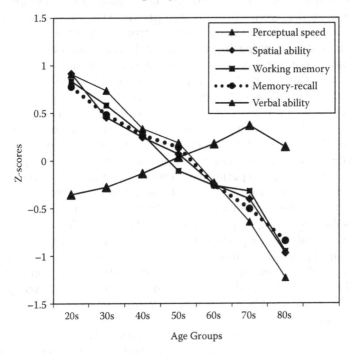

Figure 3.1 Example of cross-sectional comparisons of cognitive abilities. (From Park et al., 1996. *Psychology and Aging*, 11, 621–637.)

in prior experience with computers or the Internet. Also, assume that a comparison was conducted between younger and older adults regarding preferences for training formats and the older adults preferred face-to-face training formats and the younger adults preferred online learning. This difference in preference would not reflect a change that is implicit with aging, but rather a difference that is likely due to generational differences in prior educational experiences. There are myriad examples of generational preferences available today, such as reading e-books or the news online as opposed to hardcopy books or the printed newspaper, or texting or tweeting someone instead of calling them on the telephone. These examples will likely change in the next decade when texting will be replaced by some other type of communication.

Interindividual variability also occurs within an age group. As noted in our discussions of the aging process (e.g., Chapter 2), there is tremendous variability among older adults, within an age group. For example, if one examined reaction time or working memory among a group of 65-year-old adults one would find tremendous variability in performance. Older adults also vary tremendously in educational and health status, literacy, culture/ethnicity, skills, abilities, and of course life experiences, and these differences contribute to the vast interindividual variability seen within an age group. Also, one almost always finds that younger individuals vary less from each other than do older individuals. When comparing younger and older adults on a variety of performance measures (e.g., measures of speed or working memory), one typically finds that older adults do less well than younger people. However, as noted above, in these studies data are typically pooled within a group and the results are based on performance averages. There is also tremendous variability in performance such that some older people perform at the same level or better than younger adults and some perform more poorly and at lower levels. We compared the ability of older adults to perform a simulated computer-based customer service task for a health insurance company and found that, on average, the older adults performed less well than the younger adults (as shown by the solid regression line in Figure 3.2). However, there was also tremendous variability in performance, represented by the dots around the line in Figure 3.2, such that some of the older adults outperformed younger adults. Generally, what this means is that prediction of performance becomes less certain as people increase in age and that age may not be the best variable to use when attempting to predict someone's performance.

Intraindividual variability is that which occurs within an individual. This type of variability is often examined in longitudinal studies where changes in individuals are measured across two or more measurement occasions. In longitudinal studies, these measurement occasions are

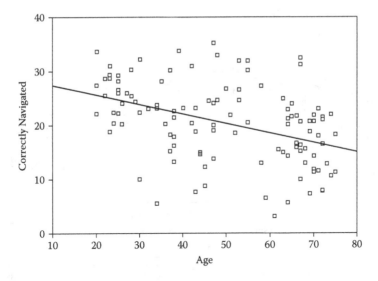

Figure 3.2 Customer inquiries correctly navigated on a simulated customer service task as a function of age. (From Czaja et al., 2001. *Psychology and Aging*, 16, 564-579. With permission.)

typically spread over years, where the performance of the same group of people on variables such as cognitive abilities is measured every 5 or 10 years. Figure 3.3 depicts changes in cognition in a single group of older adults that were measured over time. In this figure, the x-axis represents changes in age of a single cohort; individuals were measured in their twenties and then repeatedly over the years until their late eighties. In contrast, in Figure 3.1 the x-axis represents differences measured among different people in different age groups. Measurement of intraindividual differences can also occur over days or between sessions. For example, as we discuss in Chapter 11, one might evaluate a trainee's knowledge on a topic prior to training, midway through a training course, at the end of the course, and again six months after training. Intraindividual differences among older adults may reflect differences in performance due to the aging process such as changes in visual acuity or processing speed that occur from young adulthood to old age, disease (e.g., dementia), or learning and experience (e.g., world knowledge or language skills). However, there can also be short-term intraindividual variability where an individual's performance fluctuates across days or even within a given measurement occasion due to factors such as fatigue, acute illness, distractions, or attentional lapses. Think of highly trained professional athletes who play well at the beginning of a game and then lose their focus or become tired and play poorly during the middle or at the end of a game. There are a variety of factors that can contribute to intraindividual variability.

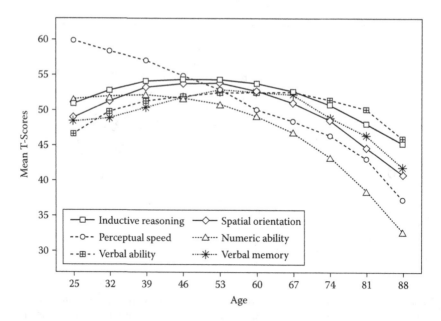

Figure 3.3 Longitudinal data on age-related changes in cognitive abilities. (From Schaie, 1996. Intellectual development in adulthood: The Seattle Longitudinal Study. New York: Cambridge University Press.)

3.2.2   Factors influencing aging and learning

There are numerous factors influencing the ability of older adults to learn new skills and procedures, or acquire new knowledge. As reviewed in Chapter 2, older adults experience changes in many cognitive abilities such as working memory, perceptual speed, and attention (see also Figures 3.1 and 3.2), which are important to learning and create challenges for older adults, especially if the task is complex or in an unfamiliar domain. For example, it may be more difficult for an older person to learn if the to-be-learned material is presented rapidly or if too much information is presented at once, given age-related changes in information-processing speed and working memory. Generally, the rate at which older adults can transfer or consolidate information that is being learned to long-term memory is slower with age. Experience is also an important factor that influences how effectively someone can learn new information. If the new learning material is related to existing knowledge, learning will be more efficient, especially if analogies are made within the training program between the new material and someone's existing knowledge. Studies of age and job performance have also shown that job knowledge is a better predictor of performance than age or performance on standardized tests of abilities.

There are also age-related changes in the sensory/perceptual systems that have an impact on learning. With age, sensory and perceptual systems become less acute, which has vast implications for the design of instructional materials. If, for example, training manuals or the text in online training programs are in small-size print or the contrast ratio is low (e.g., blue on black) older people may have difficulty reading them and may miss important information. It is also more difficult for older adults to learn if the lighting is inadequate, the room is noisy, or there are many interruptions or distractions.

Other individual differences such as health status, education, language, and acculturation also have an impact on learning. People who have a chronic or acute illness generally have less reserve capacity and thus may become fatigued more easily and need shorter training sessions or more rest breaks. Individuals whose first language is not English or who have limited education may also experience challenges in training situations. Other studies have shown that participation in training and development activities is also influenced by factors such as anxiety and self-efficacy and perceived benefits of participation. Individuals who believe that they are capable of improving and learning new skills are more likely to participate in training activities, as are those who perceive a potential benefit from participation.

Social factors may also influence how well older people learn. We found that older adults tend to prefer small as opposed to large group formats, and learning with peers as opposed to mixed groups of older and younger adults. We have also found that people prefer to learn with others who are at the same skill level. We recently evaluated a basic computer/Internet training course that was designed for older adults and delivered in community centers by community trainers across four cities in the United States. When asked for input with respect to how the class should be improved, several of the participants commented that the skill levels of the trainees were too mixed, such that people with less background knowledge felt that they were given insufficient practice on basic constructs, whereas those with more knowledge indicated that the course should have included more advanced topics. The availability of social support is also important. People who have support from management or family or friends are more likely to take advantage of training or retraining opportunities.

All of the factors discussed in this section have implications for the design of training programs. However, as stated repeatedly throughout this book, this does not mean that older adults cannot learn new skills or gain new knowledge. People of all ages can learn if they are motivated, feel comfortable in the learning situation, and receive the proper type of instruction. In fact, even older adults with mild cognitive impairments are capable of new learning. We also know that many training techniques are effective for older adults. However, there is not yet an adequate research base to determine whether some training techniques are specifically more beneficial for people on a consistent basis. It depends on the topic, setting, and the target population.

Thus, an important aspect of the design of any training program is conducting an analysis of the needs, skills, and characteristics of the training population, as well as an analysis of the tasks/activities to be trained and the environmental setting where the training will take place. For example, assume that you were designing a training program to teach a small group of seniors at a senior center how to use a spreadsheet application. In addition, you were hoping to have them use computers to complete practice exercises during training and for a post-training evaluation. In this case it would be important to determine, for example, if the seniors had any prior computer experience. If not, you would need to teach them basic computer/mouse and windows skills before proceeding to the spreadsheet training. It would also be important to gather other information about the training population such as the ability to speak and understand English, presence of sensory/motor impairments, reason for taking the class, and so on. In addition, you would need to do an analysis of the spreadsheet application to determine the skills that needed to be taught to perform basic tasks. Finally, you would need to determine if the senior center had the necessary amount of available computers and other training tools such as a whiteboard and if there were adequate lighting, space to accommodate the number of students, and minimal distractions. We provide general guidelines for a number of these issues later in this chapter.

3.2.3   Brief synopsis of the aging and training literature

There have been a number of studies examining learning and older adults. This literature includes studies examining age differences in laboratory tasks such as learning and subsequently recalling a list of words. What is typically found in cross-sectional studies of younger versus older people for these types of tasks, is that on average older people do less well on most indices of performance such as number of words recalled from a word list. There is, of course, variability in performance in all age groups, but especially among older adults as we discussed earlier in this chapter.

There have also been meta-analyses of older adults and job training. A meta-analytic study gathers the results from a large number of studies that address a common set of research topics or hypotheses, and attempts to determine the average *effect size* for a particular research question or issue. Effect size refers to how big, on average, is the difference *d* in performance between the two groups. In this case, a meta-analysis might be examining age differences in job training outcomes (e.g., performance on a training assessment test) and the average size of the difference between younger and older trainees. Effect sizes are usually given in standard deviation units for the mean difference and are usually categorized, in the social sciences, as small, medium, and large when d is .2 units, .5 units, and .8 units, respectively. The other statistic that is often used in meta-analysis is the correlation coefficient r, which indicates the degree of association between two variables such as age and response time. For r, examples of small, medium, and large effects would be values of .10, .30, and .50.

Meta-analyses comparing age differences in job training outcomes generally show age differences for the most part favoring the younger trainees. For example, older adults generally show less mastery of the training material on immediate post-training tasks when compared to younger adults, and generally require more help and hands-on practice. They also typically take longer to learn and to perform evaluation tasks. This is not surprising, as one of the most established findings regarding aging is that older adults (typically those in their sixties and seventies) take about one-and-a-half to twice as long as young adults (those in their twenties) to perform any new task. General slowing is seen in all kinds of activities, both cognitive (learning) and physical (response time). Some of the slowing may be attributable to older adults' preference for accuracy over speed, a type of speed/accuracy tradeoff. For younger adults, the opposite is often true; in other words they may trade off some accuracy in their responses in order to respond more rapidly.

Studies have also examined age differences in abilities to learn specific skills or tasks outside the work setting. For example, given the explosive use of technology in almost all domains (Chapter 1), our group and many others have examined if older adults are able to learn to use new technologies or technology applications such as a text-editing or social-networking program. These studies encompass a variety of technologies and applications and also vary with respect to training strategies. The influence on learning of other variables, such as attitude toward computers and computer anxiety, has also been examined. Studies also vary in terms of setting and include laboratory studies that systematically vary training parameters such as training approach or pacing, and evaluation of training programs delivered in community settings such as basic computer/Internet training courses.

Overall, the results of these studies indicate that older adults are, in fact, able to use technology such as computers and the Internet for a variety of tasks. Consistent with findings discussed earlier regarding job training, older adults typically take longer to acquire new technology skills than younger adults, and generally require more help and hands-on practice. In fact, in our evaluation of the community-based computer/Internet training program, one of the most prevalent responses regarding needed improvements in the course was the need for more interactive practice and assistance from the course instructor. Findings have also shown that cognitive abilities such as working memory, processing speed, and reasoning influence success at learning technology skills such that people with higher cognitive abilities tend to do better. This is not surprising, as using technology and software applications places demands on these cognitive abilities. This is not to suggest that people with lower abilities are unable to learn; instead, they just may require more time and more practice or assistance.

It is important to note that older people, even those in the later decades, are interested in learning to use new technologies. They are not technophobic. The SeniorNet program, whose mission is to provide older adults training so they can successfully use computer applications and the Internet, has trained more than one million older adults since its inception in 1986. Currently SeniorNet has about 50 learning centers throughout the United States and also offers online training programs. However, although older adults are very receptive to receiving training on new technologies, it is not uncommon for older people to report more anxiety and less self-efficacy about their ability to learn to use technologies such as computers or smartphones than younger adults. This is important, as our data clearly show that computer anxiety and computer self-efficacy are important predictors of technology adoption. Also, attitudes toward technology and comfort using technology are influenced by experience and the nature of interactions with computer systems as well as system design. Not surprisingly, the data indicate that older adults who have a positive perception of factors such as usability, ease of use, and usefulness of technology applications, and a positive initial interaction with technology, have more positive attitudes and greater technology efficacy and are more willing to learn to use technology applications such as the Internet. This underscores the importance of designing training programs to accommodate older learners. In this regard, the National Institute on Aging (NIA) has developed a training curriculum, "Toolkit for Trainers," for instructors who teach older adults to search for and find reliable online health information. The ToolKit includes lesson plans, handouts, and practice exercises and is based on cognitive aging and vision research (see www.nihseniorhealth.gov/toolkit).

3.3 Designing training programs for older adults: General guidelines

There have been numerous attempts over the past several decades to find training techniques that are best suited to older learners. The expectation is that one form of training would prove differentially better for older adults as compared to using the same methods for younger adults. Very few instances of interactions between age and training technique have been found. For example, a meta-analysis (Callahan, Kiker, and Cross, 2003) examining the benefits of different training techniques for learners over age 40 did not reveal any consistent age and training method interactions. The methods examined were lecture, modeling, and active participation (discovery learning). All three proved effective with older adults. The meta-analysis also compared instructional factors: materials, feedback, pacing, and group size. Only self-pacing and group size (smaller is better) proved important, with self-pacing resulting in more positive outcomes for the older adults. An important caveat is that the analysis was restricted to people age 40 and above, so that the chances of finding moderator effects of age were somewhat limited.

In an early study, we compared the use of a goal-oriented interactive training approach to a passive step-by-step training approach in terms of effectiveness of teaching younger, middle-aged, and older adults how to use a text-editing program. The goal-oriented training approach was based on learning to perform certain tasks such as typing a letter as opposed to simply learning procedures and rules. It also incorporated analogies to familiar concepts in an attempt to capitalize on existing knowledge. Overall, we found that the goal-oriented approach improved the ability of our older participants to learn the software. However, there was no age-by-training interaction; the goal-oriented approach also resulted in better training outcomes for the younger and middle-aged participants (Figure 3.4). Recently, a few investigators have shown interactions with age (greater gains for older adults) for procedural (*action*) versus conceptual training for automated teller machines and for web search training. In these studies procedural training seemed to be more efficacious for older learners.

It has also been suggested that e-learning methods that incorporate multimedia formats (see Chapter 10) might be particularly beneficial to older learners, as these types of methods allow for self-pacing and flexibility with respect to information presentation (e.g., text, voice, animation).

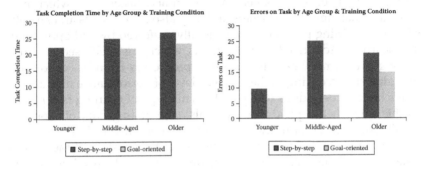

Figure 3.4 Text-editing performance as a function of age and training strategy. (From Czaja et al., 1989. *Behaviour & Information Technology,* 8, 309-319.)

However, currently there is no solid evidence to support the notion that these types of formats are particularly beneficial for older adults. We recently completed a study where we compared two introductory online training programs—a multimedia (text, narration, and animation) training format and a unimodal (text-only) training format—on the ability of middle-aged and older adults to use the Medicare.gov website to solve problems related to health management. We also included a *cold start* no-training condition, typical of how most people encounter these websites. Overall, our data indicated that training had a positive impact on participants' ability to use the website compared to the no-training condition, but there was little difference between the two training formats. However, the sample in our study was relatively small and the training was limited to a brief introductory program. Not surprisingly, we also found that prior knowledge about the Internet and cognitive abilities were important predictors of performance. This finding reinforces the importance of ensuring that people have competence with respect to basic skills, in this case mouse and window skills, before introducing them to more advanced skills such as using Internet applications such as the Medicare website.

In summary, a wide variety of training techniques is effective for older adults. There does not seem to be a particular training method that is consistently differentially beneficial for older people. As noted earlier, it depends on the target population and the task or skill to be trained. However, there are general guidelines for optimizing training programs for older adults that are based on the cognitive aging and training literature. We now provide a brief overview of these guidelines.

One basic tenet with respect to training older adults is to ensure that sufficient time is provided for them to process information and to complete training exercises and the training evaluation tasks. The pacing requirements of a training program are critical for older people given age-related declines in speed of information processing. In this regard it is typically not advisable to mix younger and older adults in training programs, as younger learners generally learn at a faster pace, which may not be adequate for older learners and may also create stress and anxiety. Self-paced training programs are generally preferred by and optimal for older learners. Adaptive training formats, which are often used in online training programs, where the training progresses from easier to more difficult tasks according to the learner's level of performance may also be beneficial for older adults (Chapter 10). This type of training is generally feasible only when applied to individuals rather than groups of people. In any situation (group or individual), if the to-be-learned task is complex, it is always important to ensure that the learner has mastery of the basic or easier levels of the task before proceeding to more advanced or complex levels. In this case the learner should also be provided with an example of the whole task at the beginning of training to help them integrate the various steps or constructs and understand the training goal. It is also advisable to provide opportunities for the learner to be actively involved in the learning process by creating an engaging learning environment that captures the attention of the learner. Passive learning situations should be avoided.

Providing the learners with feedback on the quality of their task performance at a level that allows them to understand how and why task performance was not correct is especially important to older adults. Otherwise, they may have to engage in unlearning things that are incorrect, which enhances the likelihood of confusion and increases the cognitive load. It is also important to be careful with the amount of information or feedback presented to a learner to avoid potential problems with information overload. Both positive and negative (although constructive) feedback should be provided during training. Feedback can be immediate or delayed and the frequency of feedback can also vary. With older adults, it is essential that adequate feedback be provided during the early stages of learning, and this feedback should be immediate. However, providing too much feedback is not a good idea as it may disrupt the learning process by causing learners to focus on the wrong things or overloading working memory. Thus, there is often a fine line between providing repeated or excessive feedback that is overtaxing and providing insufficient feedback that can hamper the learning process. In this regard, it is also important to ensure that help is available, easy to access, and that the learner is acquainted with sources of help.

As noted earlier in this chapter, it is also important to facilitate linkage of the new information with information that is already learned. For example, in our early study of training techniques for text editing we found that drawing analogies between familiar concepts such as file folders and file cabinets was helpful in terms of explaining new concepts. Another effective training strategy to help enhance the retention and transfer of newly learned material is to increase the amount of practice and to provide backup or simple reminder materials that can be used outside the training environment. The provision of support materials is especially important for skills or procedures that are used on a more infrequent basis, specifically those that might be used under duress such as using a healthcare device. One question that often arises is whether massed or spaced practice protocols should be used. Generally the available data suggest that for a wide variety of tasks, especially those that are complex, distributed practice is more advantageous with respect to skill acquisition and retention (Chapters 4 and 5).

Variability is also key to fostering the retention and transfer of training. During training, trainees should be provided with and given practice on a wide variety of examples of a task or skill (Chapter 8). This necessitates identifying a wide variety of contexts in which the to-be-learned task might be performed. If possible, opportunities should also be provided for the learner to encounter examples that are different from those used during training. This might involve having them complete a variety of examples in homework assignments.

It is also important that training and support materials are well designed and easy to use. Important things to consider are the size and contrast of text; pacing, amplitude, and frequency of narration; organization of material, literacy demands, and technical difficulty level of the content. We always recommend pilot testing training and assessment materials with a representative sample of trainees. Finally, it is important that the training environment is free from distractions and excessive noise, the trainees' work areas are set up using existing ergonomic guidelines, there is adequate lighting, and the temperature is comfortable.

3.4 Recommendations

In this chapter we have provided a brief overview of current knowledge about aging and learning and training, and factors that influence the learning process. We have also presented some general guidelines for the design of training programs. However, the development of a training program should be a systematic and iterative process and should involve samples of representative trainees and instructors. At a minimum, the instructional design process must consider the task, the characteristics of the trainee population, time, cost, and available resources. A summary of the recommendations discussed in this chapter is presented below.

- Older adults are interested in engaging in training and retraining programs.
- Older adults can learn new skills but they may need more time, as well as more practice and support.
- Older adults are not technophobic and are interested in learning about and using new technologies.
- Aging is associated with increasing heterogeneity. Older adults vary tremendously in educational and health status, literacy, culture/ethnicity, skills and abilities, and of course life experiences.
- There are two types of performance variability: interindividual variability in performance and intraindividual variability in performance. Interindividual variability is the type observed between people or between groups. Intraindividual variability is variability that occurs within an individual, either across tasks or measurement occasions.
- Numerous factors influence the ability of older adults to learn new skills or relearn skills, including person factors (e.g., past experience, health status), task factors (task complexity), training factors (e.g.,

pacing), social factors (e.g., social support), and environmental factors (e.g., lighting, noise).

- Many training techniques are effective for older adults. However, the type of training that works best depends on the topic, the setting, and the target population.
- The pacing requirements of a training program are critical for older people. It is important to ensure that sufficient time is provided for them to process the instructional information and to complete the training exercises and the training evaluation tasks.
- Ensure that the trainee has mastery of the basic or easier levels of the task before proceeding to more advanced material or complex levels of a task.
- Opportunities should be provided for the learner to be actively involved in the learning process.
- Both positive and constructive negative feedback should be provided during training.
- Facilitate linkage of the new information with information that is already learned.
- Provide practice on a wide variety of examples.
- Provide sufficient support during training and make the learner aware of sources of support that are available post-training.
- Training and support materials must be well designed, easy to use, and adhere to existing guidelines for older adults.
- Adequate rest breaks should be provided.
- The training environment should be free of distractions and attention should be paid to noise, lighting, temperature, and workstation setup.
- Prior to designing a training program it is important to conduct an analysis of the needs, skills, and characteristics of the training population; the tasks or activities to be trained; and the environmental setting where the training will take place.
- Prior to implementation it is advisable to pilot test training and training evaluation materials with a representative sample of trainees and instructors

Recommended reading

Czaja, S.J. and Sharit, J. (2009). Preparing organizations and workers for current and future employment: Training and retraining. In S.J. Czaja and J. Sharit (Eds.), *Aging and Work: Issues and Implications in a Changing Landscape.* Baltimore: Johns Hopkins University Press, 259–278.

Rogers, W.A., Campbell, R.H., and Pak, R. (2001). A systems approach for training older adults to use technology. In N. Charness, D.C. Park, and B.A. Sabel (Eds.), *Communication, Technology and Aging.* New York: Springer, 187–208.

Schaie, K.W. and Willis, S. (2011). *Handbook of the Psychology of Aging*. London: Academic Press/Elsevier.

Smith, M.C. and DeFrates-Densch, N. (2009). *Handbook of Research on Adult Learning and Development*. New York: Routledge.

Swezey, R.W. and Llaneras, R.E. (1997). Models in training and instruction. In G. Salvendy (Ed.),*Handbook of Human Factors and Ergonomics*, 2nd ed., New York: John Wiley, 514–577.

chapter four

Learning and skill acquisition

4.1   Learning

4.1.1   Stages of learning

When people attempt to learn a topic or material that is reasonably complex or involved, learning is not expected to be instantaneous. Such learning generally takes time and requires some degree of study, rehearsal, practice, or review. This process, whereby the learner acquires knowledge and skill associated with the topic, has often been characterized as progressing through a number of stages.

As discussed later in Section 4.2.5, researchers have also characterized the processes underlying skill acquisition as progressing through stages that correspond very closely to the stages associated with the broader spectrum of learning that is described in this section. Although the topics of learning and skill acquisition overlap in many ways, it is important to note that they also have distinctive features. Most notably, skill acquisition typically focuses on the relatively well-defined performance of some task, that is, on the acquisition of a defined skill. Learning, in contrast, is more general. For example, although some may consider learning about how a hailstorm develops to be a skill, instruction on this topic is generally considered to ensure that people have an adequate understanding of, or have learned this material. Thus, the topic of learning can be viewed as more general in nature and, to some extent, encompassing the topic of skill acquisition.

Although the precise number of stages of learning, how these stages are labeled, and what occurs within these stages have been the subject of debate, we describe a general three-stage process that can be applied to a wide range of learning situations. However, despite this broad characterization of the learning process, it will still reveal challenges that many older learners are likely to face.

Obviously, in a process as complex as learning it is not realistic to conceptualize the progression through stages of learning as discrete leaps across sharply defined boundaries. Rather, the transition across stages is much more indefinite, with elements characteristic of one stage still likely to be lingering as processes associated with subsequent learning stages become manifest. This same characterization of stages associated with the general process of learning also applies to descriptions of stages associated with skill acquisition.

An understanding of what transpires during various stages of learning can provide insight into training guidelines, and ultimately into the prospects of successful learning through the timely incorporation of different instructional strategies (Chapters 8 and 10). In addition, although the success of learning can be affected by how such strategies are dynamically adjusted or adapted to the different stages of learning, we also need to be aware of the learner's information-processing capabilities (Chapter 7) that can affect the success of learning through each of these stages. There are many other factors that can have an impact on learning, such as motivation, anxiety, fatigue, and self-efficacy (Chapter 6), but we consider the learner's information-processing system and associated cognitive capacities as crucial factors governing learning; therefore, this topic is given separate and relatively extensive treatment in Chapter 7. Also, as emphasized in Chapter 8, factors related to cognitive capabilities and limitations can, particularly for older adults, influence the application of instructional design strategies during the course of learning.

4.1.1.1   Stage one

In the initial stage of learning, the learner is generally exerting a significant degree of cognitive effort in an attempt to understand a topic relevant to performing some task. This effort is directed at understanding various stimuli such as buttons, displays, menus, switches, input control devices, spatial configurations, a host of terms, facts, concepts, and in some cases, at successfully manipulating objects. One advantage during this early stage of learning is that the information that is being processed can be modified relatively easily. Thus, faulty initial assumptions or incorrect facts can be abandoned somewhat readily without a major processing penalty because the effort that has so far been invested has not been sufficient to result in stable (i.e., hard-wired) knowledge structures that would require considerable mental effort to modify or even discard.

However, this advantage comes at the cost of the substantial investment of mental effort needed to interpret the new information that is being confronted by the learner. The learner directs this effort mostly at identifying various stimuli; discriminating these stimuli from one another by identifying differences in their properties; and making relevant associations between predominantly concrete stimuli, as when certain stimuli become associated with certain other stimuli, and discriminating between these associations as well.

During the initial stage of learning, depending on the nature or extent of the learning topic, significant cognitive effort may also be directed at understanding concepts. This requires more sophistication in the discrimination between variations in properties of objects so that the learner is able to group, in some meaningful way, collections of stimuli or objects that may otherwise be different, either in terms of appearance or other properties. For example, Internet information-seeking activities may result in the user coming across a variety of different search boxes. Some of these search artifacts may be designed to search specific documents, some may search a website, and still others may search larger or different kinds of domains. The ability to differentiate between these different objects, understand their associations with other objects (e.g., different virtual spaces), and generalize these different objects to the more abstract concept of search are all activities that may comprise basic learning.

In some learning settings, especially for many work tasks, an overview of the overall task, job, or problem domain may also be imparted at this stage of learning in order to provide the learner with a context for understanding why component tasks or activities need to be performed. For example, in training an older worker to perform a task involving the processing of insurance claims, it may be useful for the learner to understand the importance of reliable and timely processing of claims and how errors in such processing can be harmful to the work organization. Although this overview places additional demands on information-processing activities, we have found that older adults often benefit from being provided with the broader picture as long as it is kept relatively brief. Providing a sense of purpose to the task can also serve as a motivational influence, which should not be discounted, as such factors can possibly compensate to some degree for cognitive declines that older people may be experiencing (Chapter 6).

In general, trainers should consider providing overviews as they can provide a context for the more specific task-related information that will be presented, and thus help organize this information. An overview provides conceptual placeholders, so to speak, into which the learner can assimilate material. However, with older learners, and especially with technological material with which the learner may be very unfamiliar, too much preliminary information can backfire as it may not be retained by older adults as easily. In addition, it could induce confusion and ensuing anxiety, which can amplify any anxiety regarding the learning situation, perhaps due to perceptions of low self-efficacy that may have already been present. Thus, it is probably best to keep such overviews brief and simple, assuring the older learner that although the learning task may appear daunting, its mastery will be achieved systematically. Generally, the goal of the overview should be to facilitate an increased focus in attending to the task-related information.

Regardless of how the learner obtains instruction during this initial stage of learning (e.g., one-on-one, through a facilitator, in a classroom setting, or through some computer-based arrangement such as e-learning), some degree of practice should be undertaken on tasks that will need to be performed. The traditional classroom type of instruction, whereby a large amount of declarative knowledge concerning facts and concepts is imparted to the trainees prior to their performance of the tasks requiring this knowledge, generally imposes excessive demands on attention and memory, and can undermine motivation. This strategy is thus likely to be even more inappropriate for older adults. Instead, the emphasis should be more on a *learning-while-applying* approach that allows the person to access information while they are doing the task.

During this initial stage of learning, the learner is still processing new information and is likely being challenged by the need to differentiate between the various facts, basic rules, and concepts being confronted. Consequently, task performance may be unstable and characterized by slow improvement, depending on how difficult the task is and how much relevant prior experience or knowledge the learner has. The learner thus may still be taking steps backward in order to find an appropriate sequence of activities that lead to a successful outcome, or may need to rethink a term or concept or what constitutes an association or rule. Toward the conclusion of this stage of learning the learner is likely to have a basic understanding of the requirements for successful task performance, although there still may be gaps in this understanding that prevent smooth performance.

It is a good strategy to try to induce extra effort by the learner early on in training. This can be accomplished, for example, by instilling some more difficult exercises or tasks for the learner to perform. Inducing extra effort in older learners can, through the rehearsal activities that are activated, promote better retention of this information (Chapter 5). It may also help enable the material to be applied to different contexts (Chapter 8). When applying this strategy to older learners, the trainer (or instructional program) should be careful to ensure that the older learner has built up sufficient confidence, that can be inferred to some extent from the learner's performance on prior tasks or exercises, in order to prevent possible frustration from undermining the prior training experiences. That is, the sequence of skills that have been acquired needs to be carefully assessed (Chapter 8) before attempting to induce extra effort through more challenging or harder problems.

As an illustration of inducing extra effort, when training a person to perform a scheduling task (e.g., scheduling parts to be processed in a production operation or scheduling elective surgeries for hospital patients), the instructor can introduce several unique types of conditions that the scheduler may face. For example, a number of machines that process parts in a production plant can simultaneously break down or several physicians scheduled to perform surgery can suddenly be taken ill. The learner may not have as yet accumulated the requisite skills to handle these problems smoothly, however, they promote within the learner new avenues of thinking about the task elements and thus can help both transfer information to long-term memory and improve the possibility of more reliably accessing information from long-term memory (Chapter 7).

Finally, it is recommended that older learners be subjected to over-learning during early stages of training on certain basic procedural task components such as using a mouse or scrolling on a computer screen. Given the reduced information-processing capacities of older adults, this strategy provides a greater opportunity for providing them with spare capacity for managing the more difficult aspects of the task. Similarly, the designers of training programs or instructions need to pay careful attention to any issues related to legibility or identification of information, and the confusion of facts and concepts. If a reference is being made to a button on a device, the identification of this button and the legibility of any associated text or pictures should be clear, the specification of the purpose of this button should be unambiguous, and the distinction between this button and others should be apparent. Neglecting these design-related considerations can disrupt the momentum of early learning and thus seriously impair learning by older adults.

4.1.1.2 Stage two

In the second phase of learning, the learner is presumed to have attained sufficient command of the task background information, facts, and concepts to shape this information into *schemas*. These schemas are essentially packets of various forms of related knowledge that become associated in memory (Chapter 7). Schemas enable more efficient use of learned information and thus allow for smoother performance of the task. A simple example of such a schema is the *IF–THEN rule*, which has sometimes been referred to as *proceduralized knowledge*. Historically in the area of artificial intelligence, this type of knowledge encompasses what has been referred to as *production rules* that represent the elements or steps comprising how a task becomes accomplished. The use of such rules is common and can accommodate many different types of learning, for example: "If lab tests A and B are negative and lab test C is positive, then an ultrasound is recommended to rule out the possibility of medical condition X," or, "If I am in math mode in this computer application or device, then I can access functions that allow me to perform computations of type X and Y."

The linking and combining of such rules in meaningful ways is what allows many kinds of tasks to be accomplished and goals to be attained. To meet task-specific or situation-specific goals, the variables that constitute proceduralized knowledge often need to become instantiated by substituting constants in their place, for example, when substituting 160°C for the variable "temperature." This type of rule-based knowledge also can form an important basis for the many skills and strategies that people rely on during a higher stage of learning to pursue solutions to new problems.

Accomplishing such tasks, however, requires that people have knowledge structures that are sufficiently developed.

At this stage of learning, the person is likely to possess surface feature and task-specific knowledge structures about a task domain. Surface feature structures, which are typically characteristic of novice workers, are composed of the explicit and physical features of the task domain. The concepts that may comprise rules are still very fundamental and thus provide the person with the capability of reasoning about the task domain in only the most basic manner. In contrast, knowledge structures at the task-specific level allow for more complex cognitive task performance. This is due in large part to the more sophisticated nature of the concepts comprising the rules in these knowledge structures, as it is well known that the manipulation of concepts is a cornerstone of learning and discovery. Some degree of decision making, inferential reasoning, and generalization or extrapolation is therefore possible, and many workers can be considered reasonably competent at their tasks without progressing past this level of expertise.

It is important, however, to note some distinctions related to procedural knowledge, particularly between the idea of procedural knowledge in the form of rules, and procedural skill. Having conceptual knowledge, in the form of rules or in some other form, does not ensure the ability to demonstrate procedural skill, that is, the ability to perform a procedural task. Conversely, being able to perform a procedural task (e.g., speaking a language) does not imply having the underlying conceptual knowledge (e.g., an understanding of the rules of grammar).

Generally, knowledge during this stage of learning, for example, in the form of rules, may not necessarily be well-formed in the sense that the rules may not be abstract or hierarchical in nature. Consequently, when faced with a problem-solving or a complex cognitive task, only that knowledge relevant to a narrow or limited part of the problem is likely to be elicited. The knowledge is not yet organized in a sufficiently sophisticated manner to allow for the broader, more global aspects of the domain to be comprehended and integrated into the particular problem. Some of the rules may still be too general, or a schema may be underdeveloped in terms of the degree of richness and interconnectivity of knowledge. When exceptions to the rule or the schema are encountered, for example, in a practice example during training or in further accumulation of information during performance of a work task, these confrontations serve to trigger refinements to schemas. This could allow a general rule to become transformed into one or more specific rules, for instance, "If symptoms A, B, and C are present, the person likely has disease X, but if symptom R is also present, this likelihood will be reduced markedly."

The process of clarifying and solidifying declarative knowledge is thus a somewhat uneven unfolding process that depends in part on what experiences or examples the learner encounters, the ability for the learner to detect these exceptions, and the need for making adjustments to concepts and schemas. This fine-tuning process can be viewed as a bridge between the second and third stages of learning and, depending on the learning topic, may be more descriptive of stage-three learning processes.

The theme in this second stage of learning, namely that of integrating elemental learning building blocks into larger skills and knowledge components capable of providing more powerful and efficient performance, can also be manifest as a refinement in strategies. For example, in training a person to perform scheduling tasks in a production process, the more global perspective to the task that evolves during this stage of learning may allow the trainee to begin to test scheduling strategies that consider maximizing the quality of the products being produced in addition to ensuring that they are being produced on schedule.

In Chapter 8, in the context of methods of instruction, the need for learners to perform holistic tasks from the outset is emphasized, even if these tasks are very simple, as opposed to the presentation of abundant amounts of fragmented task information. Although it may seem that this kind of instructional strategy is incompatible with the stagewise conceptualization of the learning process, in fact it is not. First, there are many different kinds of learning tasks, and it is not likely that every form of learning will benefit from the initial presentation of holistic tasks. Second, even these holistic task approaches must acknowledge that supportive information—that is, some type of information needed to initiate performance—must be provided, and this information will, at least early on, be in the form of declarative knowledge or very basic procedural skills.

Finally, and perhaps the key point in a stagewise learning perspective, is that learners, and especially older learners, are hostage to their information-processing capacities. An instructional strategy may emphasize the need for presenting holistic tasks, but the learner still has to filter any task-related information through a limited information-processing system. For unfamiliar knowledge domains, this filtering is largely a bottom-up process that tries to identify and understand key declarative knowledge before it can link these knowledge elements into larger constructs.

4.1.1.3 Stage three

Depending on the task being learned, the third stage of learning may demand the longest practice period. For relatively complex tasks that call for a reasonably high degree of skill, improvement in performance is much more gradual during this stage as compared to the previous stages, and may require long sequences of repetitions. The rules developed in the previous stage become more automatic in the sense that the learner uses less conscious effort to determine whether a particular rule matches a particular situation. Likewise, the procedures or strategies developed in the preceding stage are modified and fine-tuned to increase their reliability and efficiency.

With increased learning and expertise, new rules can also be developed by combining previously learned rules into more complex higher-order rule forms. As noted, these rules may result from the learner confronting more difficult or novel problems that require reasoning at a higher or more global level, or the ability to abstract concepts and to think about these concepts (which may be an object or process) in different ways or at different (hierarchical) levels. Highly skilled or expert performance in complex cognitive task domains is often characterized not only by having a deeper conceptual understanding of the domain, but also by the ability to organize concepts and principles in increasingly abstracted ways, with a hierarchical characterization of abstracted concepts representing one type of organizational structure. Schemas may also undergo transformations where changes in concepts may trigger associations with new knowledge elements, including rules, into more sophisticated schemas or mental models (Chapter 7).

Presumably, these kinds of manipulations in knowledge can be accomplished because the learner has refined the ability to discriminate between facts as well as concepts, and to apply inductive reasoning abilities to infer general principles from facts and rules. Fundamental to these skills, and to higher learning in general, is the maturation of the learner's ability to regulate the learning process. Learners need to invoke cognitive processes that trigger, for example, chaining rules into higher-order rules, modifying schemas, or abstracting concepts in order to solve new problems. This type of regulatory maturation in learning reflects what is sometimes referred to as *metacognitive* processes or skills or knowledge about what one knows that can be used to regulate one's cognitive activities.

The achievement of *automaticity* in application of rules or schemas and the ability to concatenate rules and derive higher-order generalizations through inductive reasoning are representative of information-processing activities that are likely to challenge the cognitive processing capabilities of many older adults. Consequently, managing this stage of learning during training of older adults requires ensuring that older learners receive sufficient and varied practice on the tasks or problems associated with the learning topic.

These stages of learning are briefly summarized in Figure 4.1. In the remainder of this book, as various topics related to training and instruction and their implications for older learners are covered, there is usually no explicit reference to these stages. The processes that underlie these stages and the nature of the transitions that occur across the stages are, however, presumed.

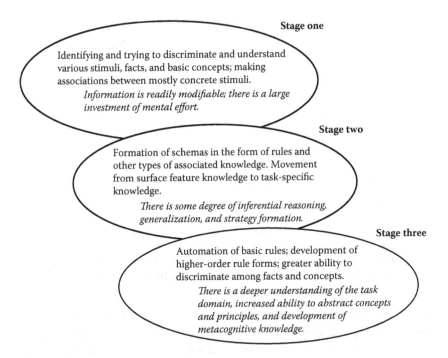

Stage one

Identifying and trying to discriminate and understand various stimuli, facts, and basic concepts; making associations between mostly concrete stimuli.
Information is readily modifiable; there is a large investment of mental effort.

Stage two

Formation of schemas in the form of rules and other types of associated knowledge. Movement from surface feature knowledge to task-specific knowledge.
There is some degree of inferential reasoning, generalization, and strategy formation.

Stage three

Automation of basic rules; development of higher-order rule forms; greater ability to discriminate among facts and concepts.
There is a deeper understanding of the task domain, increased ability to abstract concepts and principles, and development of metacognitive knowledge.

Figure 4.1 Summary of the three stages of learning.

4.2 Skill acquisition

4.2.1 A general framework

Skill acquisition can be defined very fundamentally as *the process by which people progress from lower to higher levels of performance* on some task (Charness, 2009). Obviously, there will be many factors that are capable of influencing this process, including age. The simplified framework illustrated in Figure 4.2 shows that the route to skill acquisition progresses from the more distal motivational influences to the more proximal influences of cognitive abilities, with a feedback loop from skill acquisition to motivational influences.

4.2.2 The motivation component in skill acquisition

Implied in Figure 4.2 is the decision to acquire a skill. As indicated on the left-hand side of this figure, this decision requires that the individual be sufficiently motivated to engage in the learning process underlying acquisition of the skill. The topic of aging and motivation is a complex one, and is treated in more depth in Chapter 6. However, some individual differences related to personality traits that are capable of mediating motivation

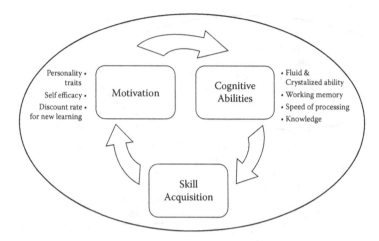

Figure 4.2 A framework for skill acquisition. (From Charness, *Aging and Work: Issues and Implications in a Changing Landscape.* Baltimore, MD: Johns Hopkins University Press, 232–258, 2009.)

toward learning are worth noting here. One such trait that could have important implications for the acquisition of skill by older adults is *openness to experience,* which appears to diminish with age. Another behavioral tendency often attributed to the aging process that could adversely affect the older adult's motivation, and thus skill acquisition, is increased rigidness and less flexibility.

The motivation or willingness to learn can also be affected by the belief that one can accomplish the task for which learning or instruction will be needed, or what has been referred to as one's sense of *self-efficacy.* The impact on skill acquisition of people's *general self-efficacy,* which is the feeling that people have about their ability to accomplish things in general, will usually be of less significance for specific learning endeavors than the more specific forms of self-efficacy. This type of self-efficacy is associated with specific learning challenges, such as one's computer efficacy with regard to learning a computer application. Charness (2009) has proposed that these and other factors related to motivation determine a person's *discount rate for learning,* which relates to how an individual perceives the value of an asset, such as learning or the acquisition of a skill, within some time frame in which the skill will be obtained and used. This topic is discussed further in Chapter 6.

When assessing one's willingness to engage in learning a new skill, people are likely to perform some type of mental cost-benefit analysis. Such an analysis would attempt to weigh the resources, in time and effort, that would need to be expended, against the value to be gained by the acquisition of the skill. Obviously, the older adult, in the face of diminishing resources (and particularly in fluid cognitive abilities, as discussed in Chapter 2 and below), may view the costs associated with the acquisition of particular types of skills as greater compared to the assessments of these costs by their younger counterparts. The effort, and thus cost associated with skill acquisition, is also likely to be affected by evidence for an increased cost of learning with aging due to adverse changes in the brain with age, with older adults often taking about twice as long to learn as younger adults. Sharit et al. (2004), in a study using subjective cost-benefit analyses for different technology devices, found that older adults rate the costs of having to learn to use and relearn to use a device (following periods of its nonuse) more heavily in cost–benefit assessments than do younger adults. Such findings suggest that future benefits of any new learning task will tend to be discounted more steeply by older adults, possibly due in part to their recognition of the greater initial (and ongoing) cognitive cost required to acquire the skill.

In addition, as emphasized in Chapter 6, given the recognition by older adults that their future horizons are more limited (i.e., that they have less time to benefit from developing the new skill), they may have less motivation to pursue acquiring skills that they believe are not worth the costs of acquisition. However, if the older adult deems the benefits of the skill acquisition to be sufficiently high or weighted more strongly than the costs, then the possible impact of *temporal discounting*, whereby the older adult may attach less value to acquiring a skill due to the lower perceived future (temporal) return on the learning investment, could be decreased. Perceived benefits may be even more important determinants of technology adoption for older adults than perceived costs.

4.2.3 The cognitive abilities component in skill acquisition

Referring again to Figure 4.2, and assuming that the older adult is motivated to learn, the efficiency of learning, which affects the rate of skill acquisition, will depend in part on the individual's cognitive abilities and current knowledge base. In Chapters 2 and 7, we discuss, in the context of the well-documented age-related declines in various cognitive abilities, an important distinction between two broad categories of cognitive ability: *fluid abilities* and *crystallized abilities*. Fluid abilities broadly reflect abilities that are involved in new learning or problem-solving performance such as perceptual speed and memory span, and after peaking in the 20s or 30s gradually decline with increasing age. In contrast, crystallized abilities, such as verbal fluency or perhaps even reasoning ability, remain relatively stable or increase throughout the life span.

As noted, aging generally affects these abilities in different ways, having mostly negative effects on fluid abilities such as perceptual speed and working memory (Chapter 2), which are believed to represent cognitive operations required for being able to solve novel problems. However, in the case of crystallized abilities such as lifelong knowledge and verbal ability, aging often has positive effects until late adulthood, as was illustrated in Figure 2.5 (Chapter 2). In contrast, Figure 2.6 (Chapter 2) illustrated a marked decline in fluid ability with increasing age. The relatively low correlation between crystallized and fluid abilities in this sample ($r = .37$) suggests that these abilities, which have been shown to be predictive of skill acquisition in new tasks, change across the lifespan in distinct ways.

Broad abilities such as crystallized and fluid abilities actually represent composites of more basic abilities that comprise critical constructs in models of human information processing (Chapter 7). To acquire a specific high-level skill, efficient use of knowledge structures and cognitive processes specific to that skill domain will be needed. One such fluid ability component, *working memory*, is especially important for the kinds of cognitive processing required for skilled performance. As discussed in much greater detail in Chapter 7, working memory (WM) is a type of workbench of consciousness responsible for both storing and manipulating information, either perceived from the environment or accessed from one's long-term memory. Working memory has been conceptualized as comprising two separate limited capacity storage systems: one manages storage of verbal information represented in verbal form (whether seen or heard) and the other stores pictorial (i.e., visuospatial) representations of information. A limited capacity *central executive* process is believed to be responsible for coordinating and managing these two subsystems.

An example of a WM test that typically shows substantial age-related decline is *alphabet span*. In this test, a list of words is presented and the person is asked to recall them in alphabetical order. This requires not only holding the list of information in memory, but also processing it further to rearrange and then report the items in the correct order. As illustrated in Figure 4.3, this measure of WM is quite sensitive to aging. This result is in contrast to the lesser slope revealed in Figure 4.4 for the short-term memory test known as *forward digit span*, which requires people to recall a string of digits in the same order as they were presented. Thus, memory abilities such as WM, which are characterized by manipulation (i.e., processing) of information in addition to its storage, are more likely to have an adverse impact on skill acquisition of higher-order cognitive tasks by older adults.

4.2.4 General slowing

Another very important concept with implications for aging and skill acquisition is the idea that with increased age the overall rate of

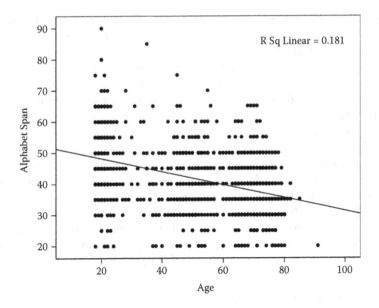

Figure 4.3 Alphabet span test score by age (yr), $n = 1{,}202$. (Data from the project described by Czaja et al., *Psychology and Aging*, 21: 333–352, 2006.)

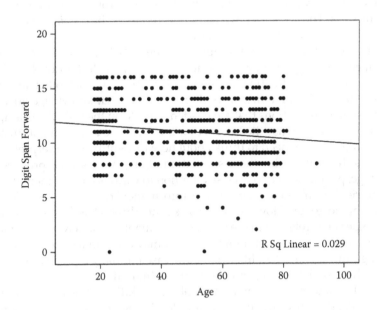

Figure 4.4 Digit span test score by age (yr), $n = 1{,}202$. (Data from the project described by Czaja et al., *Psychology and Aging*, 21: 333–352, 2006.)

information processing slows. This phenomenon can account for much of the age-related variance in higher-order cognitive processes, for example, those used for reasoning and problem solving. Analyses performed by Jastrzembski and Charness (2007) suggest that the degree of slowing by older adults relative to younger adults (i.e., the *slowing factor*) will vary depending on whether the task involves primarily motor processing (by a factor of 2.1), perceptual processing (by a factor of 1.8), or cognitive processing (by a factor of 1.7).

Thus, although knowledge, as reflected in tests of crystallized ability, tends to increase with age, processes supporting the acquisition of new knowledge tend to decrease, particularly those processes involving a slowing in information-processing speed, which can affect the rate of skill acquisition. This is a very important fact to consider in training and instructional programs, including worker training in industry, and its implications are manifest throughout this book in a number of the recommendations that are made regarding training older adults.

4.2.5   The skill acquisition process

At the start of this chapter, a general stagewise process governing broad-based learning of a topic or skill was presented. It was also noted that the underlying processes with regard to the related topic of skill acquisition are also typically characterized as progressing through a similar stagewise process. These stages have been referred to as the *cognitive phase*, the *associative phase*, and the *autonomous phase*.

Closely paralleling the description of the initial stage of learning, the first stage of skill acquisition is dictated primarily by declarative learning, where processes of attention are slow and consciously controlled, and the focus is on the often-inefficient recall of mostly factual knowledge governing the specific operations that are needed to perform the task. For example, someone learning to use the Internet for the first time to search within and across websites may need to think about how to identify and interpret various cues that will enable the instantiation of an online search process. In this stage, people need to pay close attention to various environmental cues and retrieve from their memory the information regarding the relevant aspects of those cues. However, if the task is not difficult and the learner has prior experience relevant to the task's requirements, relatively little instability may be evident and rapid improvement may occur in task performance, although this is much less likely to be the case for many older learners.

In the associative stage much of this factual information, background knowledge, and general instruction about a skill becomes *compiled* and *proceduralized* with practice. For example, the person learning to search online may now identify different organizational features of a home page and respond accordingly to rules or procedures governing what activation of these structures is likely to produce. The development of a complex skill could result in the formation of hundreds or even thousands of such rules, referred to as *production rules*. Performance in this stage is still very much under the conscious control of attention, is much more fluid, and is characterized by a much steeper learning curve than was evident in the earlier stage of skill acquisition. Toward the end of this stage procedures are fine-tuned through various mechanisms, including the use of discrimination and the strengthening of associations between stimuli and responses, and performance begins to level off.

In the autonomous stage in skill acquisition, one sees gradual improvement after considerable practice. This extensive practice period shifts control from conscious voluntary processes of attention, to low effort or automatic control of performance: what is often referred to as the shift from controlled to automatic processing (Chapter 7). The various components of the procedure thus become *automatic* in the sense that many of the relevant activities comprising the skill can be carried out with minimal attention devoted to their monitoring. That is, procedures become finely tuned to specific patterns in the environment (or what were referred to earlier in the context of learning as *higher-order* procedures). For example, in the case of motor skill acquisition, whole-word motor patterns become possible for a skilled typist: upon seeing a fairly complex word or even pairs of words, the finger movement pattern for the required keystrokes operates automatically. This stage, and the entire stagewise process of skill acquisition for that matter, can be applied to tasks characterized as much more cognitive in nature as well. For example, upon viewing a results page from a keyword search within a website, the online searcher may now be able to distinguish the features of these results from those that might be produced by a search engine's keyword search. In addition, execution of the procedure is faster.

4.2.5.1 Learning curves

The concept of a learning curve deserves some discussion given its standing in the skill acquisition literature. *Learning curves* are mathematical functions that relate outcome measures, such as time or errors in performance, to the number of practice trials or amount of practice time. These curves have been used to characterize and predict the nature of skill acquisition for a range of tasks. A variety of functions have been used to fit skill acquisition curves, including power, exponential, and logistic (s-shaped) functions, although most skill acquisition functions appear to be described by power functions. However, although a power function may provide a general estimate of skill acquisition, there are many parameters that can influence skill acquisition and these curves may not predict skill acquisition for a number of complex tasks. Although learning curves historically have occupied an important place in the area of skill acquisition, we have chosen not to review the many studies that have used learning curve functions to describe skill acquisition. Instead, our primary focus is on higher-order cognitive aspects of the learning tasks underlying training and instruction. For older adults, normative changes in cognitive ability give rise to tremendous variability within this population, thereby negating the predictive virtues of learning curves for all but the simplest of proceduralized tasks.

4.2.6   Skill acquisition and aging

A number of researchers have examined the skill acquisition processes of younger and older adults to determine how similar or different they are. For example, younger and older adults might be compared on how well they acquire a mathematical skill such as squaring two digit numbers mentally or whether older adults could learn to use and deploy mnemonics to help them remember long lists of words in roughly the same way as do middle-aged and younger adults. Generally, such results tend to indicate that older adults are less efficient than younger adults at acquiring such skills. There is even evidence (for some tasks) that following training of older and younger adults on a novel procedure, younger adults will continue to outperform older ones following continued practice by both groups over an extended period of time.

As noted in the description of the phases underlying the skill-acquisition process, the automation of processes underlying skill acquisition represents a critical higher-level stage. However, with regard to tasks involving visual search and memory search processes, there is evidence that older adults may not automate their search procedure as effectively as younger adults on visual search, which entails *consistent mapping* between cues and actions, although they do automate the process for memory search. However, even when older adults have demonstrated the ability to achieve automation of mental procedures they tend to respond more slowly than younger adults, suggesting that general slowing with age may ensure that even older adults who are considered experts at performing some task may take longer to respond than equivalently skilled younger adults.

Being slower, however, does not necessarily translate into a more negative training outcome, as in many task scenarios pure speed can be less meaningful than quality of performance. For example, analyses based on the Australian workforce demonstrated a net cost savings for training older workers (Brooke, 2003), which was mainly attributable to lower turnover rates for older as compared to younger workers. In other cases, slower performance, for example, in communicating with prospective customers when employed in sales positions, could be offset by better communication skills, the perception of being more knowledgeable or trustworthy, and ultimately a more favorable performance outcome.

4.2.6.1   The role of knowledge in skill acquisition
and skilled performance

For complex problem-solving tasks that could require searching a large space of possibilities for a solution, the role played by fluid ability skills may be overshadowed by domain-specific procedures and knowledge that could drastically curtail the number of possibilities, and thus the information-processing load, required for task performance. Ways in which prior knowledge can compensate for age-related declines in abilities have been demonstrated in the domains of chess and bridge (Charness, 1987), puzzle solving (Hambrick, Salthouse, and Meinz, 1999), and tasks involving pilot communication with air-traffic controllers (Morrow et al., 1994).

The importance of prior knowledge has also been demonstrated in a study involving young, middle-aged, and older novice and experienced word processors who were to learn new word-processing software. The older adults without prior knowledge about word processing showed typical patterns of age-related decline and slowing in performance (Charness et al., 2001). However, for those with knowledge about word processing, retraining to use a new word processor took considerably less time, which implies a positive *transfer of training* effect (Chapter 5). This effect typically occurs if the cue–response relationships of the prior task are comparable to those in the new task environment. Furthermore, performance during and following training was found to be almost unrelated to age through the middle-aged people, and slightly worse for the older (in this study, those close to retirement age) adults. Although both breadth of software application experience (in a positive direction) and age (in a negative direction) predicted performance for the overall sample of study participants. Prior software experience was more predictive of performance, and an age-by-software experience interaction effect indicated that knowledge of software had a greater effect on the performance of the older adults.

However, the reverse effect has also been demonstrated. Specifically, in a study involving use of a search engine to seek information on the Internet, Czaja et al. (2010) found that prior Internet experience was the strongest predictor of performance for a sample of adults aged 60–70. This effect was present even after cognitive abilities found to be significant predictors of performance were subsequently included in a hierarchical regression model. However, for the study participants over 70 years of age, prior Internet experience was not a significant predictor. Instead, cognitive abilities and, most important, reasoning ability, were highly predictive of performance, even after prior Internet experience was accounted for in the regression model. These results indicate that having adequately functioning cognitive abilities critical to task performance could compensate for lack of, in this case, some degree of technical knowledge related to Internet search activities.

In some of the studies cited above (e.g., Charness, 1987; Hambrick et al., 1999; Morrow et al., 1994), domain-specific knowledge critical to task performance may have been sufficiently influential to override any performance issues that could have been revealed due to general slowing or cognitive ability declines. For example, in the study by Morrow et al. (1994), older pilots may have partially been able to compensate for age-related declines in memory when doing read-back of air-traffic control messages by relying on their superior knowledge of aircraft positioning information. Knowledge, however, may not always be able to compensate for age-related declines in other, for example, more fluid abilities, especially if those abilities are central to performance and there is no way to perform efficiently if these abilities are diminished. Also, continued use of these fluid abilities may not necessarily be enough to prevent their declines. In this "use it and still lose it" position, declines in basic abilities occur even for those with extensive job-related experience and who continually practice work-related activities.

In summary, older adults are more likely than younger adults to be at a disadvantage when acquiring new skills that do not draw on knowledge that may be contained in the larger and more refined knowledge bases that older people generally possess, or perhaps even on abilities, such as reasoning, that may be linked to knowledge culled from lifelong experience. Even when the learning tasks do provide the opportunity of positive transfer from prior knowledge (e.g., word processing), older adults will take more time to reach the same criteria levels of performance, or will perform at lower levels if trained for the same amount of time. Because of these greater challenges in skill acquisition, older adults are thus likely to need greater motivation in taking on these learning challenges.

4.2.7   The ability requirements perspective to learning and its relationship to skill acquisition

The tasks that people perform, especially in job environments, typically require a mixture of different kinds of abilities. These abilities generally can be categorized into sensory, cognitive, and motor abilities. To establish a categorization or *taxonomy of human performance* based on ability requirements, tasks requiring common abilities would be grouped into the same category. Factor-analytic or other clustering methods, based on intercorrelations among task or test performances, have typically been used to form the initial basis for identifying these ability dimensions.

Table 4.1 lists a number of cognitive abilities identified by Fleishman and Quaintance (1984) who, in addition, also identified numerous sensory

Table 4.1 Cognitive Abilities

Flexibility of Closure	The ability to identify or detect a previously specified stimulus configuration that is embedded in a more complex sensory field (the specified stimulus must be disassembled from other well-defined but distracting perceptual material).
Speed of Closure	The speed with which a set of apparently disparate sensory elements can be combined and organized into a single meaningful pattern or configuration (the elements to be combined must be presented within the same sensory modality).
Verbal Closure	The ability to solve problems requiring the identification of visually presented words when some of the letters are missing or scrambled.
Associated Fluency	The ability to produce words from a restricted area of meaning.
Figural Fluency	The ability to draw quickly a number of examples, elaborations, or restructurings based on a given visual or descriptive stimulus.
Expressional Fluency	The ability to think rapidly of appropriate wording for ideas.
Ideational Fluency	The ability to produce a number of ideas about a given topic (it addresses the number of ideas produced and not the quality of those ideas).
Word Fluency	The facility to produce words that fit one or more structural, phonetic, or orthographic restrictions that are not relevant to the meaning of the words.
Integrative Processes	The ability to keep in mind simultaneously or to combine several conditions, premises, or rules in order to produce a correct response.
Associative Memory	The ability to recall one part of a previously learned but otherwise unrelated pair of items when the other part of the pair is presented.
Number Facility	The ability to manipulate numbers in numerical operations (e.g., subtracting, dividing, integrating), and involving both the speed and accuracy of computation.
Visual Memory	The ability to remember the configuration, location, and orientation of figural material.
Memory Span	The ability to recall a number of distinct elements for immediate reproduction.

(continued)

Table 4.1 Cognitive Abilities (continued)

Selective Attention	The ability to perform a task under distracting (i.e., irrelevant) stimulation; the task and the irrelevant stimulation can occur either within the same sense or across senses.
Perceptual Speed	The speed with which sensory patterns or configurations can be compared in order to determine identity or degree of similarity.
Time Sharing	The ability to utilize information obtained by shifting between two or more channels of information.
General Reasoning	The ability to select and organize relevant information for the solution of a problem.
Logical Reasoning	The ability to reason from premise to conclusion, or to evaluate the correctness of a conclusion.
Deductive Reasoning	The ability to apply general concepts or rules to specific cases or to proceed from stated premises to their logical conclusions.
Inductive Reasoning	The ability to find the most appropriate general concepts or rules that fit sets of data or explain how individual items in a given series are related to each other. It involves the ability to synthesize disparate facts, to proceed logically from individual cases to general principles, and to form hypotheses about relationships among items or data.
Spatial Orientation	The ability to maintain one's orientation with respect to objects in space or to comprehend the position of objects in space with respect to the observer's position.
Spatial Scanning	Speed in visually exploring a wide or complicated spatial field.
Verbal Comprehension	The ability (for the receiver of information) to understand language.
Verbal Expression	The ability to utilize language (either oral or written) to communicate information or ideas to others. This ability does not relate to the production of ideas or the quality of the ideas, but solely to the quality of the communication of such ideas.
Visualization	The ability to manipulate or transform the visual images of spatial patterns or objects into other spatial arrangements. It requires the formation of mental images of the patterns or objects as they would appear after certain specified changes such as unfolding, rotation, or movement of some type.

(continued)

Table 4.1 Cognitive Abilities (continued)

Figural Flexibility	The ability to change sets in order to generate new and different solutions to figural problems.
Flexibility of Use	The ability to think of different uses for objects.

Source: Adapted from Fleishman and Quaintance 1984. *Taxonomies of Human Performance: The Description of Human Tasks.* Orlando, FL: Academic Press.

abilities (e.g., peripheral vision, depth perception, glare sensitivity, auditory attention, sound localization), physical abilities (e.g., static strength, dynamic strength, gross body coordination, extent flexibility, and stamina), and psychomotor abilities (e.g., manual dexterity, arm–hand steadiness, finger dexterity). Figure 4.5 shows a portion of a binary decision flow diagram used for identifying various psychomotor abilities.

Generally, specifications of abilities and their associated definitions often differ depending on the applied or theoretical domains of interest. Thus, ergonomists and human factors researchers interested in human

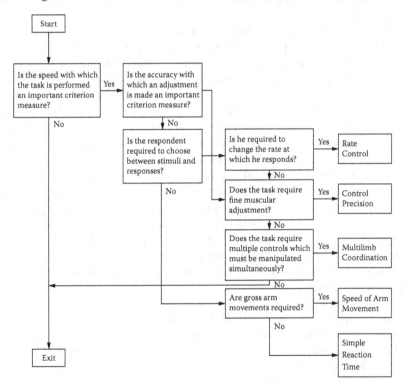

Figure 4.5 Portion of a binary decision flow diagram for ability identification and classification. (From Fleishman and Quaintance, *Taxonomies of Human Performance: The Description of Human Tasks.* Orlando, FL: Academic Press, 1984.)

work or experimental task performance may focus on or interpret certain human abilities somewhat differently than researchers in the areas of cognitive aging or neuropsychology.

For many tasks, especially those that have gradually incorporated greater degrees of automation, the fundamental abilities needed for performing those tasks have become altered. For example, psychomotor skills are becoming less predictive of pilot performance as compared to cognitive abilities (which are needed for dealing with new cockpit technologies) and verbal skills (which are needed for dealing with communication activities). Thus, it is essential for many work-related situations that a *job analysis* (an analysis of the tasks or behaviors that define a job) be performed to determine what tasks are required as part of the job. A job analysis generally includes specifying the tasks normally accomplished; specifying the environments in which the tasks are performed; and specifying the related knowledge, skills, and abilities required for successful task performance. Ultimately, *task analysis* (Chapter 8) will also likely be needed to determine the cognitive and physical requirements of the constituent job tasks.

In a landmark research study in the area of abilities and skill acquisition (Fleishman and Hempel, 1955), it was demonstrated that as people learn or gain experience in performing tasks, the contribution of various abilities to the performance of those tasks changes. Figure 4.6 illustrates some classic work in this area and specifically how the importance of various abilities changes with practice (i.e., skill acquisition) on a discrimination reaction time task. These kinds of changes with practice are likely to occur for more complex tasks as well. Thus, for some tasks perceptual skills may be critical early on in practice or training, but as the learner gains experience, reasoning skills may become more important. Overall, these kinds of findings parallel the general idea of stages of learning and skill acquisition discussed earlier in the sense that they may explain, at a more fundamental level, how expertise evolves through practice. They also imply that different tasks may tap into different abilities, so that the nature of the progression from one stage of skill acquisition to the next may depend on the types of abilities that may be required for that task as experience is gained.

In Section 4.2.4, we noted an important distinction between two broad categories of cognitive ability, namely, fluid and crystallized abilities. An important implication of this distinction and, more generally, of age-related declines in cognitive abilities, is that the ideal progression of learning may be hampered, depending on the abilities required for task performance, by the fact that critical abilities required during the early stages of performance might be subject to decline. If other more viable abilities are unable to compensate for those that are diminished in their capacity, then different, perhaps more innovative,

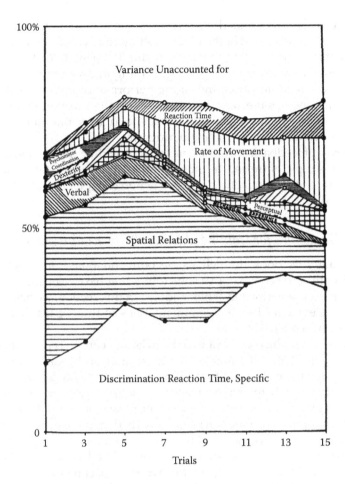

Figure 4.6 Percentage of variance represented by loadings on each factor at different stages of practice on a discrimination reaction time task. (From Fleishman and Hempel, *Journal of Experimental Psychology*, 49: 301–312, 1955.)

methods of instruction may need to be applied to these older learners in order to lessen the dependency of their learning on these diminished abilities. In Chapter 8, as well as throughout this book, we try to emphasize aspects of training and instruction that might circumvent the potential for age-related declines in abilities to impede the progression of learning and skill acquisition.

4.2.8 The feedback component in skill acquisition

Finally, the role of feedback in the skill acquisition framework depicted in Figure 4.2 should not be overlooked. For example, feedback reflecting performance failure could diminish one's sense of self-efficacy, which may affect the motivation of the older adult more adversely. Similarly, the perception that the learning process is mentally taxing may discount the value of further learning more steeply, leading to decreased motivation. On the other hand, feedback indicating performance success or the perception that one has the capacity to cope with the learning materials or process could heighten one's self-efficacy, and thus the motivation to endure in a training or instructional program.

The role of feedback is a critical factor in providing instruction to older adults. This topic is treated in more depth elsewhere in this book, particularly in reference to methods of instruction (Chapter 8) and e-learning (Chapter 10).

4.2.9 Guidelines for training older adults to acquire skills

Throughout this book, guidelines, recommendations, or insights are injected within the context of various topics that have relevance to training and instruction directed at older adults. In this section, we briefly note some of the evidence-based recommendations in this regard that can be found in Charness (2009).

One issue that has been examined is whether certain types of training are better suited for older adults. In a meta-analysis by Callahan, Kiker, and Cross (2003) of studies of training methods, notably lecture, modeling, and active participation (or what is sometimes referred to as discovery learning), for adults aged 40 and older no substantial interactions of age with training technique were found, with all demonstrating the capability for being effective modes of training for older workers. This result is not all that surprising; as we show in Chapters 8 and 10, it is how training is administered that is more likely to have an impact on the effectiveness of learning and skill acquisition than the type of training.

Other variables examined in the meta-analysis study included materials, feedback, pacing, and group size. Of these, self-pacing and group size (specifically, having a smaller group size when instructor support is required) had significant effects for training outcome, with self-pacing being the most important variable. Pacing is a variable that can directly affect the human's information-processing capability. For older adults, as emphasized throughout this book and implied above in the discussion of general slowing of information processing, applying an inappropriate pace to the presentation of instructional material is probably the surest way of compromising learning in older adults.

When training is provided on novel material in a lecture-style group format, for example, in work settings, another factor to consider is the age composition of the group. In light of the varying rates of learning in younger and older adults, the recommendation is to use an age-segregated training policy. However, if the instruction being offered to workers can be related to prior knowledge, given that past knowledge is a very strong predictor of new learning in older as well as younger adults, knowledge-segregation (e.g., into novice, intermediate, and expert groups) as opposed to age-segregation, may make sense as an approach to training and support of skill acquisition. Furthermore, it may be particularly beneficial to older adults who have prior knowledge related to the topic if the instructional techniques can draw on, or otherwise highlight, the structure of knowledge representation that the older learner possesses. This could facilitate learning by the older adult as it would impose less information-processing effort to manipulate or integrate prior knowledge resident in memory into a form that is consistent with the instructional material.

Another issue that is at the core of methods of instruction (Chapter 8) concerns how practice sessions should be distributed. A general and fundamental rule in training is that spacing out practice sessions results in superior skill acquisition for new information than massing practice into fewer sessions. For older adults, there is also the factor of fatigue to consider (Chapter 6), which could make massed practice even more detrimental for older learners. In theory, there probably exists some optimal schedule for practice, as ideally one would want to balance the possibility for decay of previously learned material (by virtue of not having a practice trial to boost its strength) against the level of activation of the to-be-learned material. For example, in massed practice sessions, there may be less concern for the decay rate but more concern for the readiness for absorbing new learning.

However, as pointed out by Charness (2009), as older adults probably have slightly different activation and decay rates for memory compared to younger adults, systematic investigation will be needed to determine the optimal spacing of practice for older adults. Also, following from the discussion on the role of prior knowledge, these activation and decay rates could be altered dramatically for older adults depending on the extensiveness of their relevant prior knowledge and the degrees to which the training method draws upon this knowledge. For older adults, with fatigue, self-efficacy, and other factors that may be looming to threaten the learning experience, it may be best to chance possible decay of information and ensure that sessions are ended with a feeling of confidence and positive expectations toward impending learning or practice sessions. The issue of possible decay across sessions, perhaps because of insufficient activation of memory traces in previous sessions, could be addressed through *refresher strategies* that seek to bridge the gap between learning across sessions by beginning each session with an overview of the previous session and the highlighting of key concepts or procedures.

4.3   Recommendations

- Provide an overview of the task domain and objectives so that the learner can have a reference or cognitive structure as a way to organize and associate new learning materials.
- Emphasize learning by doing instead of passive learning that requires the learner to retain excessive amounts of information in memory prior to performing any task or exercise that requires use of that information.
- Induce extra effort early in training by requiring performance on some relatively more difficult exercises, but only if older learners have already shown signs of having gained confidence based on their performance on previous more basic exercises.
- Consider using an overlearning strategy to ensure that the older learner is proficient on procedural components of tasks that are required to be performed repeatedly within the context of other, more cognitive task components. This helps to ensure that these procedural components will not take critical processing resources away from them as they take part in more demanding cognitive processing activities.
- Early in learning, ensure that labels, facts, instructions, and concepts are distinguishable to the older learner to eliminate or reduce the possibility for confusing these elements of learning.
- Consider the fact that information-processing activities associated with the skill being learned that involve speed, such as perceiving information on displays or manipulating information, slows with aging. These task components need to be addressed more carefully so that they do not negatively affect the acquisition of new knowledge and skills.
- Ensure that the older learner is comfortable with the pace of instruction; if not, adjust the pace accordingly.
- Avoid massed practice sessions; instead, space out the practice sessions.
- Always try to end practice sessions with experiences that bolster the confidence of the older learner, for example, by completing a reasonably challenging problem correctly or providing a correct response regarding the meaning of an important concept.
- Always ensure that gaps between practice sessions are appropriately bridged by commencing new practice sessions with at least brief refresher overviews of the previous session(s), and also possibly reinforcing key concepts or procedures from previous practice sessions.

Recommended reading

Ackerman, P.L. (1992). Predicting individual differences in complex skill acquisition. *Journal of Applied Psychology*, 77: 598–614.

Anderson, J.R. (1982). Acquisition of cognitive skill. *Psychological Review*, 89: 369–406.

Bandura, A. (1989). Human agency in social cognitive theory. *American Psychologist*, 44: 1175–1184.

Fisk, A.D. and Rogers, W. (1991). Toward an understanding of age-related memory and visual search effects. *Journal of Experimental Psychology: General*, 120: 131–149.

Fitts, P.M. and Posner, M.I. (1967). *Human Performance*. Belmont, CA: Brooks/Cole.

Kanfer, R. and Ackerman, P.L. (2004). Aging, adult development, and work motivation. *Academy of Management Review*, 29: 440–458.

Koubek, R.J. and Salvendy, G. (1989). The implementation and evaluation of a theory for high level cognitive skill acquisition through expert system modeling techniques. *Ergonomics*, 32: 1419–1429.

Melenhorst, A., Rogers, W.A., and Bouwhuis, D.G. (2006). Older adults' motivated choice for technological innovation: Evidence for benefit-driven selectivity. *Psychology and Aging*, 21: 190–195.

Newell, A. and Rosenbloom, P.S. (1981). Mechanisms of skill acquisition and the power law of practice. In J.R. Anderson (Ed.), *Cognitive Skills and Their Acquisition*. Hillsdale, NJ: Lawrence Erlbaum, 1–55.

Salthouse, T.A. (1996). The processing-speed theory of adult age differences in cognition. *Psychological Review*, 103: 403–428.

Schaie, K.W., Dutta, R., and Willis, S.L. (1991). The relationship between rigidity-flexibility and cognitive abilities in adulthood. *Psychology and Aging*, 6: 371–383.

Vazquez-Abad, J. and Winer, L.R. (1992). Emerging trends in instructional interventions. In H.D.Stolovitch and E.J. Keeps (Eds.), *Handbook of Human Performance Technology*. San Francisco: Jossey-Bass, 672–687.

chapter five

Retention and transfer of training

5.1 Retention

5.1.1 Memory systems: A key to learning and retention

The stagewise perspective to learning (Chapter 4) is just one of many frameworks that can be used to illustrate the process of learning. Instructional practitioners and even theorists often find it more convenient to view the way people learn in terms of the constraints imposed by memory. As discussed in detail in Chapter 7, there are two memory systems that dictate much of a person's ability to process information and thus learn: *working memory* (WM) and *long-term memory* (LTM).

WM is the workbench of information processing. It has limited storage capacity and high sensitivity to loss or confusion of input information if attention is not focused on activities taking place there. LTM is the repository of lifelong knowledge that is resident in many forms. It also serves as the basin for rehearsed material from WM, as well as the supporting knowledge structure for WM when thoughts, ideas, manipulation of information, or any activity is being negotiated in WM. The pivotal role of WM in learning becomes even further enhanced when one considers WM's structural characteristics. In particular, as discussed in Chapter 7, WM contains two distinct memory subsystems: one that processes spatial information (e.g., pictures) and one that processes verbal information (e.g., spoken or read words). In addition, these subsystems are subject to their own limited capacities.

As we show in Chapter 7, the limited capacity of WM is critical to understanding the link between human information-processing attributes and the ability to learn. The learning theory that perhaps best personifies the important role played by WM (Chapter 7) is referred to as *cognitive load theory* (Sweller, 1994). This theory describes three types of cognitive loads or processing activities that can influence the success of training and instructional programs: (1) essential or intrinsic cognitive load, (2) extraneous cognitive load, and (3) generative or germane cognitive load.

Essential cognitive load is the inherent level of difficulty associated with instructional materials. Because all instruction has an inherent difficulty associated with it, designers of instructional programs need to be cognizant of it. For example, learning how to formulate, construct, and analyze problems using spreadsheet software is more difficult (specifically, it embodies more difficult concepts and requires more challenging reasoning processes) than learning how to input a paragraph of text using word-processing software. This type of load, which is also referred to as *intrinsic cognitive load*, addresses those aspects of the instructional content that are directly relevant to the learning goals and thus need to be managed in as optimal a way as possible in order to promote learning. The important point is that although the inherent difficulty of the learning material cannot be altered, methods or formats of instruction (Chapters 8 and 10), for instance, may help to greatly diminish the cognitive load imposed by the learning material.

Extraneous cognitive load refers to aspects of the instructional or learning experience that compel the learner to use limited WM capacity in an unproductive manner or to direct WM resources toward material that is not relevant to the learning goals. Consequently, extraneous cognitive load can undermine learning. Because this load can be attributed to the design of the instructional materials, in principle it is under the control of instructional designers. For example, if the instruction was attempting to describe how a machine works using only spoken words, learners are forced to use their limited cognitive processing ability to retain various facts related to this information while trying to use cognitive processing to visualize the machine and how the information relates to the machine's structure and functioning. Dividing the learning information into spoken words and pictures can lessen the cognitive load considerably. Directing any WM capacity to information that does not constitute essential cognitive load, such as background music or interesting but irrelevant side notes about the learning topic would also impose extraneous cognitive load as this information is unnecessary for learning. For older adults, the presence of extraneous cognitive load when the intrinsic cognitive load of the learning material is already relatively high can greatly increase the likelihood of poor learning outcomes.

Generative cognitive load represents the deeper forms of cognitive processing undertaken by the learner that can accentuate the learning experience, for example, those that might be fostered by appropriate types of practice exercises. From the standpoint of the designer, handling this load is primarily about managing complexity, and many approaches to instructional design can be brought to bear to meet this objective. With respect to the spreadsheet example, separate holistic tasks directed at the formulation, construction, or analysis aspects of these problems can be worked on separately, enabling learners to build into their LTM chunks of knowledge, sometimes referred to as *schemas* (Chapter 7), which relate to these different aspects of spreadsheet use. The use of scaffolding, whereby the learner gradually takes on more responsibility for solving a problem through worked examples, and appropriate sequencing methods, whereby the construction of the tasks is designed to ensure reinforcement of previous material as well as extension of the learning topic, can also help ensure that the learner does not get overloaded at any stage of the learning process (Chapter 8). These approaches also enable learners to build into their LTM the requisite knowledge and skills prior to moving forward in the learning program, which facilitates further deeper learning. Many of the methods of instruction brought up in this book are discussed particularly within the context of the older learner, as these methods, if appropriately tailored to this population, can offer the possibility for greater success in learning more difficult material.

Building on cognitive load theory and the characteristics and constraints associated with WM and LTM, Clark and Mayer (2008) emphasized three fundamental processes that underlie learning, and thus which the instructional environment, whether in the form of an instructor, workbook, or computer-based technology, needs to support: selecting, organizing, and integrating. *Selecting* refers to the ability for the learner to focus attention on the relevant aspects of incoming visual and auditory information. *Organizing* refers to the transformation of the training content, such as spoken or written text, pictures, and videos, into mental representations that can become the building blocks needed to support the development of new knowledge and skills. *Integrating* refers to a higher level of coordination of verbal and spatial information with existing knowledge in LTM to attain an understanding and ultimate mastery of the learning material.

Another idea related to cognitive load theory that has important implications for the processing and recall of information, especially for older adults, is the concept of *environmental support* (ES). With ES a person can rely on aspects of the environment to reduce demands on information processing such as identifying relevant cues or recalling information that was previously available (which reduces the burden on retention of information). This idea thus extends beyond learning situations to include task performance in the more general sense. For example, many situations involving learning or task performance require that information be held in memory (facts that were just taught or appeared while working on a task) while this information is also processed (connecting these facts to previously learned concepts or to rules that one has available concerning how to perform the work task). Concurrent storing of information in memory and processing of that information places demands on limited WM capacity (Chapter 7), which can potentially be offloaded with appropriate ES. Similarly, forgetting of recently learned or processed information that is needed for further learning or task performance can benefit from particular forms of ES.

Relative to the effects on younger adults, ES in principle can improve, reduce, or have no effect at all on older adult performance depending on the particular nature of the ES, the type of learning or task situation, and individual older adult characteristics such as sensorimotor and cognitive abilities. Many of the improvements in older adult performance through use of ES result from its ability to compensate, to some degree, for declines that older adults may be experiencing in these abilities.

In the review of the concept of ES by Morrow and Rogers (2008), emphasis is given to identifying those conditions under which age differences in performance might be reduced or magnified by ES. To aid in exposing design considerations when developing ES, they offer a framework that points to two general functions which ES can provide: reducing task demands and supporting the use of the individual's cognitive processing resources (Chapter 7). Both of these ES functions can be directly or indirectly beneficial to older adults during instructional situations as these learners try to retain and recall information in support of ongoing learning. In reducing task demands, three categories of ES are considered: enhancing task-relevant information (e.g., making critical information more salient), increasing processing opportunity (e.g., slowing the pace of information presentation to allow more time for processing the information), and externalizing the task (e.g., providing placeholders or reminders to lessen the dependence on recall). In supporting the use of cognitive resources, two categories of ES are considered: encouraging knowledge use (e.g., use of metaphors that allow the learner to link prior knowledge with the learning material),and guiding allocation of resources (e.g., cueing the learner as to where to find information in an online worked example).

Recall of information can also be aided through the provision of reminders. Table 5.1 summarizes characteristics of good reminders. These criteria were originally developed for the purpose of minimizing errors related to forgetting that plague many workers who perform maintenance tasks in industry. However, given the issues related to retaining information that older adults may encounter in learning situations, a number of these principles can easily be extended to learning situations as well. For example, with regard to the *confirm* criterion it suggests that in designing online worked examples the learner should easily be able to determine which problem steps have already been performed so that extraneous cognitive load is not invested in determining what was done and what still needs to be.

5.1.2   *Retaining what was learned*

Following episodes of learning and practice through task performance, many people find themselves disengaged from the task activity for significant durations of time. Depending on how much opportunity was available to perform these tasks and the nature of the expertise the tasks required, these gaps in time may have a negative effect on the ability for a person to recall critical skills and knowledge. An important goal in any instructional

Table 5.1 Some Characteristics of Good Reminders

Universal Criteria

Conspicuous: It should be able to attract the person's attention at the critical time.

Contiguous: It should be located as closely as possible in both time and distance to the to-be-remembered (TBR) task step.

Context: It should provide sufficient information about when and where the TBR step should be carried out.

Content: It should inform the person about what has to be done.

Check: It should allow the person to check off the number of discrete actions or items that should be included in the correct performance of the task.

Secondary Criteria

Comprehensive: It should work effectively for a wide range of TBR steps.

Compel: It should (when warranted or possible) block further progress until a necessary prior step has been completed.

Confirm: It should help the person to establish that the necessary steps have been completed. In other words, it should continue to exist and be visible for some time after the performance of the step has passed.

Conclude: It should be readily removable once the time for the action and its checking have passed.

Source: Reason, J. 1997. *Managing the Risks of Organizational Accidents.* Aldershot, Hampshire, UK: Ashgate.

process is to enable the trainee to retain the learned knowledge and skills following periods of time in which the task is not performed.

Achieving this objective is particularly challenging for older adults, primarily due to age-related declines in cognitive abilities. These declines can adversely affect all information-processing activities (Chapter 7), including the retrieval of information from memory. Consequently, older persons generally can be considered to be more prone to less reliable retention of learned knowledge. Also, older people may be more likely to face relatively longer periods or gaps in time between occurrences of task performance. This could be due, for example, to their increased tendency to engage in part-time work or because of a lower frequency in engagement of certain common task activities (e.g., use of the Internet). For these situations, one basic solution for ensuring that material is retained is to increase the amount of practice, even possibly by extending the number of practice trials beyond the point at which the learner has mastered the task, that is, to the point of *overlearning* (Chapter 4).

Clearly, an important influence on the magnitude of retention is the degree of original learning, and this tenet is probably even more relevant to older adults who, due to normative neurological changes, are likely to be subject to a greater rate of decay of memory traces. Also, the decreased reliability of older adults' memory can, in turn, make it more likely that they will confuse new learning with previous learning, which could occur to a greater extent if the learning situation requires large amounts of material to be learned. Finally, the benefits of extensive practice for older adults may be particularly valuable for procedural tasks, which generally show rapid declines in retention.

Overcoming these concerns with older learners will require balancing a number of factors, some of which were discussed in relation to skill acquisition (Chapter 4). For example, because older adults probably need greater inoculation against memory decay through more extensive learning, there will be ancillary concerns for fatigue and motivation (Chapter 6), and for the capability to process the instructional material (Chapter 7). Thus, providing older adults with more instruction, both in quantitative terms and, as suggested below, perhaps also in the qualitative nature of the instruction, may require uniquely tailored instruction recipes that blend appropriate methods of instruction (Chapter 8) into a correspondingly appropriate number, duration, and spacing of practice sessions.

The quality or nature of the instruction can also have an important impact on long-term retention of learning. As discussed in Chapter 4, injecting difficult problems into the instructional process in a systematic way can promote greater degrees of learning and thus improve the prospects for retaining the learned knowledge. This approach is consistent with suggestions that call for providing the learner with opportunities for deeper levels of mental or elaborative processing of the training material as a means for increasing the prospects of reliable retrieval of that material at some later time (Chapter 8). In some tasks such as rule-based tasks, which require carefully discriminating between the input information before selecting an appropriate rule or procedure, strengthening the degree of learning would require ensuring that the learner is aware of subtle differences in contexts that could affect rule selection. In these and many other learning contexts, *task analysis* (Chapter 8) can be especially useful for designing training programs for older adults. This method can help establish which aspects of the task may be problematic in terms of generating potential confusion or other kinds of difficulty, and thus be able to identify which learning segments may require more practice in order to ensure better learning and consequently improved retention.

Another factor related to learning that has also been found to have a powerful influence on retention is the degree to which the learning material has been well organized. To some extent, the ability to categorize incoming information that is being learned is dependent on WM ability, as well as the ability to retrieve relevant information from LTM. Some of the difficulty related to retention with older learners may be addressed through presentations that enhance the organization of the underlying concepts in the learning material, which would allow for material to be cognitively correlated and ultimately for the strengthening of schemas (i.e., associated material) in LTM. Task organization in the presentation of learning material has also been shown to benefit retention of procedural tasks which are generally prone to declines in retention.

An additional factor that has been found not only to facilitate retention, but also to improve the ability for the learner to apply the training material to new situations (i.e., to generalize to situations that were not addressed during training), is to use instructional strategies that combine conceptual knowledge with procedural knowledge. This factor is pertinent to teaching work-related skills such as troubleshooting, but can also be applied to training people on other tasks, such as the use of software applications. The idea is to integrate *prescriptive information,* that is, knowledge of how to do things, with *structural* and *functional information* that helps the learner understand the fundamental concepts that underlie how something works. However, it is essential that the learner have the opportunity to gain practice on *holistic tasks* (Chapter 8) that require the integration of such knowledge, and to be exposed to a number of different kinds of practice tasks or examples that rely on such integrated skills and knowledge.

In Chapter 4, the spacing and distribution of practice were discussed with regard to their potential impact on older adult skill acquisition. These factors can also influence the retention of skills and knowledge. For example, the so-called *spacing effect* refers to the phenomenon that material be learned at widely spaced intervals if retention of the material is important. Although the mechanisms underlying the spacing effect may not be entirely clear, one possibility is that the larger intervals of time between sessions provide the opportunity for reflection and thereby for the concomitant maintenance rehearsal that is engendered, which is critical to ensuring that information is reliably transferred from WM to LTM (Chapter 7). For older adults, this type of informal rehearsal can be very valuable. It not only reinforces factual knowledge, but can also help the learner recognize where confusion exists and thus which aspects of the material may need to be emphasized or re-emphasized to a greater extent. In this sense, this type of reflection has the ability to promote the development of what was referred to in Chapter 4 as *metacognitive knowledge*—knowledge about what one knows—which can be used to regulate one's cognitive activities. One benefit of having such knowledge is that it can prompt the learner to think about how the learned material could be applied in different contexts.

The key with older adults is to motivate them to reflect during these intervals, possibly by encouraging them to think about the learned material, or by indicating that the next session will begin with their presentation of a casual overview of what was previously learned and areas that were still of concern to them. If the learner in fact anticipates this approach to training, reflection may become more internalized. In this regard, one of E. D. Gagné's (1978) general instructional strategies for enhancing long-term retention is particularly relevant to older learners: provide for and encourage elaboration of the presented material during learning as well as during the retention interval. In addition, Gagné also suggested reminding learners of their currently possessed knowledge and skills that are related to the to-be-learned material, and ensuring that presented material is repeatedly incorporated into the instruction. These strategies are also particularly important for older learners as they serve as mechanisms for ensuring that the learned material bears stronger memory traces.

In general, massed practice sessions are, as implied previously, a bad idea for older adults. From the standpoint of acquisition of skills, the literature has generally been consistent in the finding that *distributed practice* (i.e., multiple exposures to material over time) is superior to *massed practice* (i.e., concentrated exposure to the learning material in a single or very few learning sessions), long rest periods are superior to short rest periods, and short practice sessions between rests are superior to long practice sessions between rests with respect to skill acquisition. However, there has not been consistent evidence that these variables benefit long-term retention.

5.1.3   Additional considerations related to retention

A general guideline related to retention is to encourage active generation of learning as opposed to passive presentation of the material. This guideline is very much related to the role of interactivity in generating more effective learning (Chapters 8 and 10). There are many forms of active generation and, as discussed in these chapters, the learning context may dictate how active generation may strengthen or possibly interfere with learning. However, in many contexts, active generation can serve as a form of maintenance rehearsal that strengthens memory traces and perhaps also serves as a cue to think further about the material. In this respect it can be a very valuable approach to instruction of older adults. In fact, the teach-back method (Chapter 8), whereby physicians instruct their (often older) patients to repeat back to them information concerning medication or other instructions so that the physicians can be more reasonably assured that their patients understand the instructions given to them, represents a simple and potentially very effective break from traditional passive presentation of instructional material. Not only could it enhance comprehension of the instructional material through reflection, but it could also cue the type of qualification about the instructions that is needed from the instructor.

Related to the concept of teach-back is the idea of having the learner attempt to perform a new task immediately following a demonstration. This takes advantage of the *recency effect* (Chapter 7), whereby information that has been more recently thought about and moved to LTM has a greater likelihood of being recalled correctly. Of course, depending on the circumstances surrounding the demonstration, it may not be feasible for the learner to attempt the task following its illustration, nor will it always be the best approach to learning. Depending on the task, and in particular, the extent to which it is governed by underlying concepts that may be complex, some period of incubation may be preferable that encourages deeper thinking about the material, and thus better learning and recall. However, for many procedural tasks that involve recalling steps, and for which it is neither incumbent nor efficacious for the learner to engage in deeper thought processes, immediate "try-back" (analogous to teach-back) could be especially beneficial to older learners whose memory traces may be less resistant to distraction and other factors that can intervene between the stimulus (i.e., the demonstration) and any required responses.

Another very fundamental retention guideline with important implications for older learners is to relate new learning to previous experience or knowledge that the learner may have. Taking advantage of the older adult's potentially greater crystallized ability has been repeatedly emphasized (Chapter 4), especially as a means for compensating for possible age-related declines in fluid abilities (Chapter 2). Linking new material being presented to previous knowledge represents yet another strategy for promoting such compensation. This strategy also facilitates the retention, and thus retrieval of information from LTM by taking already well-established traces and schemas in memory and strengthening or enhancing them, as opposed to requiring the older learner to establish entirely new memory traces.

In many work operations, people are required to perform highly proceduralized work activities. In fact, these procedures are often formally represented, in printed or computer-based formats, as standard work procedures that workers are expected to follow. Some basic guidelines related to the ordering of the text and instructions within such procedures, which follow from the implications of WM limitations (Chapter 7), are: avoid instructions that require retention of words whose meaning is not apparent until a later portion of the sentence is read; use *congruent* (as opposed to incongruent) instructions, where the order of the words or commands corresponds to the order in which they are to be carried out; and try to avoid the use of negation as this imposes an added load in WM (and thus may be forgotten from WM as that instruction is retained). These guidelines are especially important for older workers, and extend to the more encompassing guideline related to congruence: that the overall procedure used in training should be designed so that it matches the procedures that will be used in the operational setting. As we show in regard to transfer of training, this guideline is very consistent with a fundamental guideline for promoting positive transfer of training.

It may be tempting, especially in work settings, to address possible memory challenges facing older workers through compensation strategies in the form of greater degrees of performance support such as job aids. These types of performance support can benefit all workers by removing the burden of relying on "knowledge in the head" (Chapter 7) with environmental support (that represents "knowledge in the world") at the time this information is needed. In fact, principles exist for designating those situations that would benefit more from either training or job performance aids. For example, with regard to task complexity and ease of communication, one should emphasize training for those tasks with difficult adjustments and procedures that can only be achieved through practice and for those tasks that are hard to communicate through words. Much simpler tasks and tasks that would benefit from tables, graphs, and flowcharts are much more suitable for job performance aids.

The critical question is whether this type of balance in emphasis between training and job aids neatly translates to older workers. Swezey and Llaneras (1997) provide a retention guideline that advocates not relying exclusively on job aids to reduce memory reliance. Instead, they favor a greater reliance on "knowledge in the head," which could be derived through practice during training on retrieving the appropriate response procedure from memory. For older workers, analysis of the job and task contexts should dictate which point along the internal–external knowledge continuum is most appropriate. For example, if there is concern for cognitive overload due to unanticipated situations that can arise (e.g., changes in priorities, work orders, the need to override standard procedures), then for older workers in particular it is best to resort to job performance aids (that could provide various forms of ES) for those task aspects that can best accommodate this kind of support in order to free cognitive resources for more serious control and decision-making activities.

Finally, it is worth noting that many of the factors which can help promote retention of learned material are also likely to be those that can contribute to more effective learning. This would include organizing the practice material at systematically increasing levels of detail and ensuring sufficient time for practice problems, studying worked examples, and querying an instructor, to enable the learner to digest relatively large amounts of information. The systematic manipulation of the scope and difficulty of the practice material is discussed at length in Chapter 8 within the context of the method of *scaffolding* of practice tasks. Older learners can especially benefit from the application of these guidelines, as they can greatly affect the absorption, and thus retention, of information.

5.2   Transfer of training

5.2.1   Defining transfer of training

Transfer of training refers to the way in which previous learning affects new learning or performance. When the effect of the previous learning is to improve new learning, for example, by making it easier to grasp the concepts associated with the new learning material, or making performance on the task associated with the new learning material more reliable or faster as compared to what would have been achieved if the previous learning were not provided, then previous learning is said to have produced a *positive transfer* effect. The opposite effect, *negative transfer,* is said to occur when the previous learning results in more confusion or less reliable performance on a new task than would have been produced without this prior learning.

To illustrate the concept of positive transfer, imagine a situation in which someone is trying to learn to use a software application such as Microsoft (MS) Excel. Prior exposure to a completely different MS application, such as MS Word, should enable an individual to adapt more readily to many of the features associated with the Excel application, even though these two applications are used for completely different purposes. Someone with experience with an application similar to Word but which was designed for use with an entirely different operating system would not have the basis for this positive transfer. The possibility even exists for there to be a negative transfer effect if this individual encounters confusion due to prior habits that dictated using certain strategies that the new application cannot accommodate. Negative transfer is often observed in situations where new technologies are brought in, for example, for performing x-rays in hospitals or for control room monitoring in manufacturing or nuclear power plants. In such situations, those workers with extensive experience on prior technologies may have a more difficult time adapting to the new systems as compared to workers with comparable knowledge about the work tasks but without the exposure to the previous systems. Similarly, an older adult attempting to follow instructions from a nurse on how to use a blood glucose meter that uses a radically different approach to taking measurements than the previous device this patient used may encounter more difficulty learning how to use this device as compared to an older patient who had no prior experience using a blood glucose meter.

The nature and extent of transfer of training will depend on a complex interplay of many factors, including age. The generation of positive transfer of training in older adults is obviously an important goal. If they are taught a computer application, we want them to be able to learn other applications more quickly and more effectively. If they are being taught new operational procedures at work or need to undergo job retraining, we want prior learning to increase their success at adapting to the new work tasks. As noted in the discussion on aging and skill acquisition in Chapter 4, with older adults there is the concern for rigidity and reduced capability, as compared to younger people, for learning new things or adapting to new situations. Consequently, the ability to identify learning or instructional scenarios that have the potential to induce positive transfer, perhaps because of their intrinsic characteristics, is particularly valuable for the older adult population.

5.2.2 *Measuring transfer of training*

To measure the amount of transfer of training, it is necessary to know how long it takes a target population of people without training to acquire the needed skills for the task or job. For example, a group of workers with no training can be monitored until their performance time on a collection of tasks reaches some company's standard level. This standard performance level could be based on some synthetic time prediction system, which is essentially a model that contains a taxonomy of human activities and associated parameters that affect performance. An example of a simple *performance activity* might be "reaching out with one's hand," which can be affected by the parameters "distance reached" and "care used in exercising the reach activity." The analyst, using a task analysis, specifies the activities and associated parameters to be used by the synthetic prediction system to predict performance times for a standard operator for the particular system or job being modeled (Lehto and Buck, 2008). One well-known synthetic prediction system employed in industry is the motion-time-measurement (MTM) system, which could be used to estimate the cycle time to reach a standard level of proficiency.

The degree of transfer of training is usually expressed as a percentage. In the example above, the group without training is considered the control group. Suppose this group takes 100 hours to reach the MTM (time to criterion) standard. If another group of workers considered equivalent to this control group were given X hours of training prior to performing the job and were able to achieve standard performance more quickly, for example, in 80 hours (on average), then the savings in performance time could be attributed to the transfer of training to the learning situation and computed as:

$$\% \text{ transfer} = (Z_c - Z_t)/Z_c \times 100 = (100 - 80)/100 = 20\%$$

where Z_c is the time (or performance) required on the task by the control group and Z_t is the time (or performance) required on the task by the training group. Thus, percent transfer is based on the savings in time to reach a specified performance level on some operational task (or tasks). If the training group were to take longer on the specified task than the control group, negative transfer would be exhibited, implying that the prior training somehow interfered with or inhibited the learning of the skill or knowledge required for the specified task.

Note that in this example, the X hours of training received by the training group did not factor into the calculation of percent transfer. This has led to the consideration of an alternative measure of transfer of training, the *transfer effectiveness ratio* (TER), which expresses the effectiveness of savings due to training as a function of the time spent in training (Roscoe, 1971):

$$TER = (Z_c - Z_t)/X_t \times 100$$

For the example above, if the time spent in training was $X = 10$ hours, then TER = $(100 - 80)/10 = 2$. This ratio can be interpreted as follows: for every hour spent in training, the effect of transfer of training was to provide a savings of two hours. Both types of transfer of training measures could be used to measure the time required to reach some criterion level of performance. However, the application of transfer of training expressions can also be based on other measures, such as the number of trials required to reach some criterion level of performance, the level of mastery reached after a given amount of time or trials, or the number of errors made in reaching a given level of mastery.

The TER is usually a negative exponential function of time in training, meaning that TERs diminish, at first steeply and then less so, with increased times in training. The TER can only be negative if the training group's time to criterion performance is greater than that of the control group. Also, although it may seem that TER values less than one reflect an inefficient effect of training, this implication is based purely on the amount of time. It is possible, however, for training to be effective when TER < 1 on the basis of cost or safety. For example, training involving hazardous operations using a simulator will not result in actual accidents during the training. Similarly, actions that could result in enormous monetary losses, for example, due to damage to products, could be drastically curtailed during training. These benefits during transfer to the actual tasks or jobs are not captured in formulas such as the TER.

5.2.3 Causes of positive and negative transfer of training

It is important to maximize positive transfer of training for older adults. This will likely lead to much greater efficiency with respect to their needed investment of cognitive processing resources, and thus their ability to expand their knowledge and skill sets to other tasks that they may desire or need to perform. If one can anticipate which tasks learning could be transferred to, then emphasis could be given in training to ensure that factors which can induce positive transfer to these tasks are promoted, and factors that may induce negative transfer do not come into play. This requires identifying the factors or elements of tasks that are capable of producing positive or negative transfer of training.

In a *stimulus–response perspective* to the transfer of training effect, the emphasis is on the stimuli or cues that the learner needs to identify and the actions that need to be performed in response to these cues or input signals. The nature of the transfer of training effect, which can range from positive transfer on one extreme to negative transfer on the other, would depend on how closely the relationships between the stimuli and the corresponding responses in the prior task (the task for which some degree of knowledge or skills has been obtained) match those of the stimuli and corresponding responses in the transfer task (the task to which we want this prior expertise to be transferred). Specifically, if the new stimuli are similar to the ones previously encountered and the responses to these stimuli remain the same, *high positive transfer* will result. However, if the stimuli are identical (or extremely similar) but the required responses in the original and new tasks differ, negative transfer will occur. Between these extreme points supposedly lie intermediate degrees of transfer. If the new stimuli are dissimilar from the original ones but the responses to these stimuli remain constant across the tasks, *slight positive transfer* is presumed, and if the stimuli are identical across the original and new tasks but the responses are similar, but not identical, almost *no transfer* will occur.

In the *cognitive perspective* to the transfer of training effect, more complex cognitive processes associated with elements in the training and transfer tasks, such as information coding and retrieval of information from memory, dictate the process of transfer. Consider again the example of learning MS Excel when one already has existing knowledge of MS Word. The ability to recall and extrapolate procedures used in the Word application related to cutting and pasting text to similar functions in Excel would exemplify the instigation of more complex cognitive processes, which collectively provide for the ability to predict how these procedures might become manifest in an Excel template.

The problem of predicting transfer of training effects is a complex one as there are many factors, including the learning setting, prior knowledge, information processing, and a host of individual variables that could affect the nature of transfer. Further complicating this issue is that determining the extent to which stimulus and response elements coexist in complex tasks is not always easily accomplished. Even if these task elements could be identified and adequately described, it may be difficult to ascertain the combined effect of these various stimulus and response elements. For example, training on a simulator for the purpose of controlling an industrial chemical production process may have consistency between many, but perhaps not all the stimuli and responses across the simulator and actual operating conditions. If the element for which negative transfer might be induced turns out to be critical to system performance, the effects of positive transfer on the other elements may be inconsequential.

The difficulty of predicting transfer effects is also demonstrated from findings that transfer can actually be increased by reducing the *fidelity of simulation* in training. For example, a low fidelity simulator (defined below), by virtue of having fewer stimuli to attend to, may produce more effective learning of procedures that need to be carried out during emergencies by enabling cause–effect relationships between actions and system outcomes to be more clearly discernible. Low fidelity effectively reduces the similarity between the training and transfer contexts and, according to most models of transfer, should result in negligible or no transfer. However, the use of training simulators introduces a number of additional elements into the equation that could explain the findings of increased transfer and, more generally, help predict the usefulness of simulators in training. One important element concerns the feature of fidelity, with the possibility that different degrees of fidelity may be most appropriate at different stages of learning.

For example, early in training, high degrees of *physical fidelity* (the degree to which the simulator device duplicates the appearance and feel of the corresponding equipment used in the operational context) and *environmental fidelity* (the degree to which the simulator or simulation duplicates the sensory stimulation associated with the task as it is performed in the real operational setting) could actually confuse the learner and thereby impair transfer. This is actually consistent with cognitive load theory (as discussed above). That is, depending on the knowledge base of facts and concepts associated with the learning topic, high physical and environmental fidelity can actually promote *extraneous cognitive load*. For older adults in particular, critical WM capacity could become consumed with processing information irrelevant to the kind of learning needed at that time. With further increases in skill acquisition, however, the learner can benefit from a higher degree of environmental and, to a lesser degree, of physical, fidelity. This pattern carries over to the final stages of learning and skill acquisition, where the requirements for these two types of fidelity continue to increase.

Environmental fidelity is especially important for task contexts with large perceptual or pattern-recognition components that are critical for decision making or problem solving. For example, in training pilots to handle wind shear or other types of environment-related perturbations, simulators that can faithfully capture these challenges can provide pilots with the opportunity to process these stimuli more reliably and to determine the actions needed for countering these challenges more effectively.

Another characteristic of fidelity to consider is *psychological fidelity*, which reflects the extent to which the person interacting with the simulator perceives the device as a close fit, from both the standpoints of physical and environmental fidelity, to the actual operational setting. This type of fidelity is important, as it could influence the degree of motivation of trainees in their interaction with the simulator. For example, a person can be trained to play the role of a customer service representative for an online store, such as Amazon.com, who must go online in order to read and respond to issues that fictitious customers have communicated through e-mails. In responding to these e-mails, the trainees would need to negotiate the company's customer and product information databases, as well as other online sources of information, in order to provide an appropriate response. These types of simulations emulate the jobs of part-time teleworkers who can work anytime and from anyplace (where one can get online, such as at home), and are thus potentially a very good fit for many older adults. Exit interviews with older adults who performed this simulated task (Sharit et al., 2004) revealed that a large number of them felt as if they were immersed in an actual job; the fact that many of these participants in fact inquired into the opportunity for obtaining this job position suggested that psychological fidelity was obtained. It has become easier to achieve psychological and even environmental fidelity in simulating job tasks because a wide variety of such tasks are now performed through computer interfaces.

These simulations, by virtue of the experimental control that they can afford, are very useful for training, for identifying performance issues, and ultimately for identifying design interventions. Re-creations of systems or environments through simulations are especially good training vehicles for older learners, as they enable trainers to establish the kinds of controls on pace and complexity that are compatible with older adult requirements. In addition, such simulations are efficient in the sense that a wide variety of scenarios in an optimal sequence for learning (Chapter 8) can be presented.

5.3   Recommendations

5.3.1   Retention

- For older adults, when the intrinsic cognitive load of the learning material is already relatively high, identify any possible sources of extraneous cognitive aspects (i.e., aspects of the instruction that are not directly relevant to the learning material) and remove these sources.
- To help ensure the older learner is able to retain learning material, especially when instructing on complex learning topics, consider using various methods of instruction such as segmenting the topic

into subtopics to allow the learner to develop "chunks" of related separate knowledge, and having the learner practice on holistic tasks related to these chunks of knowledge. When practicing a subtopic, have the learner gradually take on more responsibility for solving a problem.

- Design practice problems or tasks so that they ensure reinforcement of previously learned material as well as extension of the learning topic. This improves retention of previous material and minimizes the possibility of the learner getting cognitively overloaded at any stage of the learning process.
- Consider using various forms of environment support, such as a display screen in an online learning program that contains a table of definitions or a picture depicting labeled components, to minimize the need for older learners to have to recall or retain information while they engage in exercises that would require using that information.
- To help ensure that instructional material is retained, especially when the learner is expected to experience relatively large gaps in time between using the knowledge and learning it, consider increasing the amount of practice, even possibly by extending the number of practice trials beyond the point at which the learner has mastered the task. However, in doing so, adjust the number, duration, and spacing of the practice sessions to avert the problem of fatigue on the part of the learner.
- Provide the learner with opportunities for deeper levels of thinking about or rehearsing the training material as a means for increasing the prospects of reliable retrieval of that material at some later time.
- Pay close attention to the organization of the learning material to ensure that concepts or task procedures are well organized and thus lessen the possibility of learners confusing rules, concepts, or procedural steps in a task.
- For complex tasks, consider integrating knowledge about how to do things with information that helps the learner understand the fundamental concepts that underlie how something works.
- For older adults, avoid scheduling learning sessions too closely together so that the learner can have the opportunity to reflect on the learned material, which helps to rehearse and thus reliably retain the information in memory.
- Encourage elaboration of the presented material during learning and remind learners how the currently possessed knowledge and skills that they have attained to that point during instruction are related to the to-be-learned material. This can help ensure that the learned material bears stronger memory traces in older learners.
- If feasible, have the learner attempt to perform a new task immediately following a demonstration of how that task should be performed.

• Ensure sufficient time for practice problems, and studying worked examples, as well as for questions if an instructor is providing training or guidance.

5.3.2 Transfer of training

• The stimuli used to cue the learner during training (e.g., cues that can signal when to consult a dropdown menu, when to process a particular type of insurance claim, when to lower the temperature in a production control process) should be as nearly identical as possible to the stimuli that will be encountered in the transfer setting. These cues should map to consistent responses across the training and transfer situations.

• Because older adults will be more prone to confusion when slight differences in stimuli require identical responses, adequate training is essential to ensure that the older learner understands the subtle distinctions between cueing stimuli with respect to response requirements. If necessary, provide labels that could help the learner discriminate between subtle differences in features of the input information.

• Gradually decrease the dependency of the learner on prompts and guides made available during training, so that by the end of training there are none present that would be likewise absent in the actual task or job situation.

• Gradually make the practice tasks during training more similar to the transfer tasks.

• Vary the training task scenarios to increase the likelihood that the learner can handle a larger problem space of tasks that could be encountered in the operational setting and possibly even generalizing to new task situations. The use of simulation methods is conducive to providing the kind of flexibility and control for realizing these kinds of objectives.

• Particularly with older adults, do not emphasize high variability in practice tasks or examples until the learner has had sufficient practice on the prior tasks and has demonstrated proficiency on them. Otherwise, the injection of too much variability too early in training could hinder transfer of training.

• When using physical simulators or human–computer interaction with simulation software applications as vehicles for training for transfer, it is essential that the older learner be thoroughly trained in all relevant technical features associated with the device or application so that features associated with the simulator or simulation program do not become obstacles in their own right. This may entail separate practice sessions that focus exclusively on the device or application features that are distinct from the details associated with task performance.

- If the emphasis will be on transfer to an almost identical situation to the one in which training is being given, high physical and environmental fidelity will probably enhance transfer of training.
- In contrast, if the training is directed at enabling the learner to handle tasks that may need to be performed in a variety of different environments, less attention should be given to achieving high physical and environmental fidelity. Instead, the emphasis in the use of simulation should be directed more at how the simulation can be used to perform or solve a variety of different tasks.

Recommended reading

Boud, D., Keogh, R., and Walker, D. (1985). Promoting reflection in learning: A model. In D. Boud, R. Keogh, and D. Walker (Eds.), *Reflection: Turning Experience into Learning*. London: Kogan Page, 18–40.

Farr, M.J. (1987). *The Long-Term Retention of Knowledge and Skills: A Cognitive and Instructional Perspective*. New York: Springer-Verlag.

chapter six

Motivation, anxiety, and fatigue

6.1 Motivation

6.1.1 The older adult's motivational system

Motivation has been defined as an activating force or tendency that directs and amplifies, initiates and terminates, coordinates and delineates, current mental and physical functioning toward a valued outcome of action. More simply, motivation can be thought of as an individual's inclination to allocate his or her personal resources (principally in the forms of time and effort) to a given task, and is a dynamic process that can be influenced by both the person and the environment.

Adult aging exerts a variety of influences on the individual, and these influences may directly or indirectly affect the older adult's motivational system. Generally, the literature indicates that the transition from early to late adulthood is accompanied by selective losses in cognitive, intellectual, and physical abilities, gains in job-related knowledge, gains in emotion regulation skills (e.g., there may be a decrease in anger with age, which increases the likelihood of bringing about goal success), and changes in motive systems, such as those concerning achievement (action goals that primarily involve a concern with standards of competence) and affiliation (e.g., social companionship). Specifically, during early adulthood, achievement goals are most prominent and the individual's interests and activities tend to be more organized around information that promotes achievement and opportunities for the future. In contrast, during adult aging, one's goals become increasingly organized around affect, whereby individuals seek social interactions for the purpose of achieving emotional satisfaction (e.g., social companionship), and make greater use of emotion-regulation strategies.

Thus, as individuals age there is a shift in emphasis within the individual's motivational system from the engagement of primary control over the environment in order to effect changes that are congruent with one's needs and desires, to secondary control processes that are more consistent with self-protection. Sorkin and Heckhausen (2006) suggest that as people age they are more likely to pursue goals for which they are intrinsically motivated—that is, from within, and consistent with the individual's needs and motives for task- and mastery-oriented forms of achievement motivation—rather than extrinsically motivated, such as by salary or job position.

In addition to, and perhaps even tied to these more general life transitional influences, a potentially important factor that can influence an older adult's motivation for new learning is personality, and specifically, as discussed in Chapter 4, an individual's willingness to learn new things. In the personality literature, this willingness is related to the trait known as *openness to experience*. Because of the tendency for openness to experience to diminish with age, and more generally, for adults to become more rigid and less flexible as they age, engaging older people in instructional programs directed at new learning may be an uphill battle.

These motivational influences may be further compounded by specific forms of *self-efficacy*, that is, negative perceptions of one's ability to successfully tackle specific domains of knowledge or skills such as those that might be involved in learning to use a new technology. Negative self-appraisals by older adults regarding their efficacy in specific domains may be mediated in part by metacognitive factors (Chapter 4) related to perceived declines in their cognitive abilities and the lack of critical knowledge.

Another perspective to aging and motivation, and one that is particularly relevant to the issue of older adults adopting training or choosing to learn new tasks involving interacting with new technologies, relates to the phenomenon of *temporal discounting* (Chapter 4). This idea refers to the decrease in perceived value of an asset if one must wait for a period of time before obtaining the asset. The literature related to aging and discounting is mixed, and can be argued in terms of either a tendency for lower or higher temporal discounting for older adults depending on how one thinks about the motivation for temporal discounting. For example, if discounting of delayed rewards is thought of as an impulsive reaction related to instant gratification, older individuals should display a lower rate of discounting along with their lower levels of impulsivity. If, however, discounting is thought of in terms of likelihood of receiving the delayed reward, then as the age of the individual increases, long delay times (10+ years) may fall outside their predicted life spans, resulting in older individuals discounting at a higher rate for long delays.

From the perspective of aging and skill acquisition (Chapter 4), we also know that older adults learn new material (that is unrelated to prior knowledge) more slowly than younger adults. The assumption, therefore, is that such learning will likely be perceived by older adults as constituting a significant investment in effort. In addition, we also know that older adults have less time to accrue the benefits of new learning based on their higher expectation, relative to younger adults, of dying in the next time interval, implying that their perceptions of any potential gains from this learning that can be applied to their future will be less as well. Thus, a rationally behaving older adult should discount the value of any new learning more steeply than a younger adult because the cost for attaining this learning is greater and the period from which older adults could benefit post learning is shorter as compared to younger adults.

6.1.2 *Motivating older adults to learn new tasks*

Assuming this logic is reasonable, then it suggests that we need to convince older adults that learning new material or tasks will be simpler than they may foresee, perhaps because we are confident that we have developed effective methods for training them. Also, we may need to assure older learners that the effects of transfer of the knowledge and skills that they will acquire will allow them to adapt more easily and continuously to variations of the technology, device, or system that they received instruction on as such changes unfold (e.g., in the form of new versions of a product or software application). This would enable benefits, presumably now at a relatively lower investment of cognitive processing costs, always to be available sooner rather than later. In this way, making potential benefits more salient should lead to lower temporal discounting of the value of new learning by older adults as they would become more willing to invest time in new learning.

There are thus two fundamental assumptions: (1) older adults will initiate new learning if they can be convinced that the instructional method will be relatively painless and that tangible gains can be realized soon (i.e., the investment in effort will result in a high present worth); and (2) that this initial learning will lead to progressively less investment in future efforts directed at incorporating new knowledge and skills for performing newer tasks (e.g., using the same basic technology). Although these assumptions have not been empirically tested, they appear reasonable.

What implications might these perspectives on aging and motivation have for training and instruction of older adults? First, we have to assume that there are many reasons why older adults could benefit from instruction. For example, many technological products increasingly are providing functionalities that foster self-preservation and social engagement, mostly through their ability to provide enormous amounts of information and the capability for social networking.

Also, we should not underestimate the many older adults who may be intrinsically motivated to master various tasks or material, whether it is in the languages and arts, or work tasks that could provide additional income on a part-time basis while also satisfying feelings of self-worth through the perceived contribution of the work to society. With many older adults leading longer and healthier lives, and with engagement in productive lifestyles linked to better health, there may be increased motivation to master various tasks and devices.

The idea that older adults are not interested in learning is a myth; their desires to learn may just be based on a different motivational system, one that is not as centered on attaining a goal for the purpose of primary control of the environment. In addition, there are many more reasons today for older adults to want to learn or receive instruction. Information and computer-based technologies offer opportunities for older adults to empower themselves in ways that were previously unthinkable, whether it be grocery delivery, management of health, finding needed services, increased ease of use of appliances and home environmental control systems, personal safety and security, socialization, games such as bridge, languages, or learning topics that they had wanted to learn but were never able to due to competing pressures.

In fact, one could argue that many older adults may have even more motivation to engage in learning and instructional activities than their younger counterparts. Improved and sustained health and well-being in many older adults can fuel this motivation, and having more time, without the pressure that usually exists in earlier adulthood to earn a livelihood, possibly support a family, and related activities, can more easily provide for its instantiation. The relatively recent recognition of the importance of leisure, which was once characterized as some residual realm that one entered into after having performed basic life necessities such as earning a livelihood, but which is now regarded as a significant source of well-being, is also consistent with the potential desire for older adults to pursue interests through learning. Staying productive or occupied, whether through work-related or leisure activities, may maintain or improve intellectual and physical functioning and quality of life, and may be protective or forestall various detrimental health-related consequences.

Healthcare in particular is a domain that can provide a strong motivation for learning by older adults. Learning to use the Internet to search for health information can enable older adults, who typically have more chronic illnesses than younger people and more frequently require healthcare services, to easily and conveniently obtain additional information that could have positive impacts on critical decisions, such as trade-offs associated with various surgical or drug intervention procedures, that they need to make for themselves or people for whom they are caring. This knowledge could also serve to improve the interactions that these older adults have with their physicians. From smart phones and personal health records to medical devices and telemedicine systems, numerous devices and systems that can provide various forms of healthcare information and services are continuously being developed and marketed, which could greatly improve the quality of life for many older adults.

In summary, despite tendencies related to aging that could adversely affect the motivation for older adults to learn new things, there are also compelling reasons why older adults may have the motivation to engage in later-life learning, although the basis for this motivation is likely to be different than that which drives younger adults. At the same time, as with younger adults, one should expect large individual differences in motivation. Just as some younger adults are content not to invest the effort needed to advance in jobs or take on new pursuits, there will be older adults who may rationalize that the effort required for new learning is not worth it, especially at their stage of life. However, as long as older adults perceive benefits associated with learning or instruction, whether they are tangible or less tangible in nature, there is little reason to doubt the concomitant motivation to pursue these learning opportunities.

6.2 Anxiety

For many people, uncertainty is a harbinger of anxiety, and it can be argued that this response stems from a natural part of our evolution. Organisms need to be more alert in uncertain situations, and greater arousal in these conditions serves a protective function. This type of arousal or response to stress can, however, be counterproductive when the uncertainty that we confront is not potentially threatening, at least physically, and can lead to mental paralysis.

For example, when people such as workers in hazardous environments who are responsible for controlling complex systems or students taking an important exam confront highly unexpected problems whose solutions have important consequences, there are tendencies to engage in behaviors that reflect a lack of fluidity in processes of attention and thinking. Typical kinds of errors that are committed under these circumstances include: *thematic vagabonding*, where one jumps from issue to issue, but never pursues any theme to its natural conclusion, often picking up previously abandoned attempts, forgetting that they were already pursued; and *encysting*, where the person invests excessive attention to a small aspect or a few details of the problem, often at the expense of other, perhaps more relevant information.

Learning is a classic example of a situation where anxiety can become manifest due to the prospects of needing to attend to, comprehend, and retain potentially new facts, concepts, and procedures. The anticipated anxiety and concomitant stress response will be highly variable across individuals, and may depend on the nature of the learning or instructional situation. In particular, it may be affected by how unfamiliar the material is expected to be (i.e., the degree of uncertainty), what people know (metacognition) about their own abilities, and the implications of failure to learn the material.

The experience of stress may manifest in the form of emotional, physiological, or behavioral responses that could include negative self-evaluations, increased state anxiety, and increased levels of cortisol and other physiological measures such as heart rate, blood pressure, and perspiration. The concern here is that such elevated stress levels could affect cognitive functioning as well. In fact, high stress levels have been found to impair cognitive ability and performance on various types of cognitive tasks, including those involving memory and attention. A possible explanation for this effect is that high stress levels result in a less effective allocation of cognitive resources to cognitive tasks.

If this is the case, then older adults may be more vulnerable to high stress levels than younger adults due to declines in processing speed and more limited available cognitive resources to perform a complex task. There is also a classic work which demonstrated that older adults can become more aroused, with regard to cardiovascular reactivity, relative to younger adults when coming in for an experiment (Eisdorfer, Nowlin, and Wilkie, 1970), possibly reflecting anxiety regarding the anticipated task or lower feelings of self-efficacy regarding their memory relative to their younger counterparts.

The possibility also exists that stress can result in a positive effect on cognition, particularly for cognitive tasks that produce moderate levels of stress. The well-known Yerkes–Dodson law (Yerkes and Dodson, 1908), which describes an empirical relationship between arousal and performance, dictates that performance increases with physiological or mental arousal, but only up to a point. When levels of arousal become too high performance decreases, but mostly for cognitively demanding tasks in which divided attention, working memory, and decision-making processes can become impaired. This process is often illustrated graphically as an inverted U-shaped curve (Figure 6.1). These tasks may require a lower level of arousal in order to facilitate concentration. Because research has shown that different tasks require different levels of arousal for optimal performance, this curve would not be accurate for tasks that involve, for example, simple discrimination or tasks demanding stamina. These kinds of tasks may be performed better with higher levels of arousal, which may serve to increase motivation.

Often we hear people speak of the need for more adrenaline to get going or to sharpen up one's senses and attention. Generally, trainers and instructors do not want to see learners with arousal levels so low that the learner is essentially disaffected. In such states, learners may not be willing or able to invest the attention needed to absorb new material. However, instructors must also be wary of hyped-up learners whose arousal levels are so high that there is concern that they may be unable to integrate, discriminate, or more generally, make sense of the incoming information,although there is little worry about them attending to the stimuli.

In general it is fair to assume, for several reasons, that older adults in learning situations will be more likely to experience anxiety and be more vulnerable to the possibility of being over-aroused and thus subject to some degree of cognitive impairment deriving from this anxiety. First, older adults are likely to have lower self-efficacy than younger adults. This

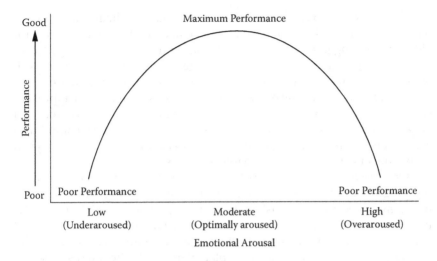

Figure 6.1 The common version of the Yerkes–Dodson law depicting an inverted-U relationship between arousal and performance for primarily cognitively demanding tasks.

may stem from the self-recognition of declining fluid cognitive abilities, which are crucial to being able to grasp concepts quickly, retain them, and integrate them with other relevant material. Second, depending on the material, if the learning situation involves concepts related to new technologies or information that the older adult is less likely to be as familiar with as compared to younger adults, older adults are likely to experience greater trepidation. Third, if the instructional situation involves groups of people, as might occur in certain work environments, or if there is perceived time pressure involved in the instruction, the older learner can be expected to experience greater anxiety, especially if the consequences of failure (e.g., not being retained for a job) are severe.

One technique that may benefit older people in complex learning situations to reduce stress is relaxation. The benefits of relaxation for cognition may be to enhance attentional capacity and help consolidate newly processed information in memory. Dijkstra et al. (2012) investigated the potential benefits of two interventions: a relaxation intervention and a problem-focused coping strategy (a type of strategy that focuses on finding different workable solutions for the problem at hand) on a sample of 119 younger adults (mean age = 19.3 years) and 108 older adults (mean age = 72.6 years) who were recruited to learn to use a PDA and a webcam, neither of which the participants had experience with. The effect of the combination of the two interventions was also examined. Data were collected on physiological and subjective levels of stress, mood, anxiety, and workload, and on task performance. The interventions, particularly the coping-only and the relaxation-only interventions, were found to help reduce physiological and subjective arousal and, in several cases, to a greater extent for older than younger adults. The older adults tended to start the experiment with more apprehension than their younger counterparts but, as a result of being subjected to relaxation and coping procedures prior to performing the task, they may have been better able to cope with this apprehension by feeling less aroused physically, and experiencing less anxiety and negative affect. The intervention combining relaxation and coping also contributed to higher performance on the webcam task but did not yield higher benefits than the separate interventions.

One lesson that emerges is that instructors and designers of training programs, including those responsible for designing online training programs, need to give consideration to the potential anxiety that older populations of learners may be harboring. Otherwise, the likelihood increases that much time and effort invested in training will be forfeited. This concern is similar to the "blah blah blah" phenomenon discussed in the context of limitations in cognitive capabilities in Chapter 7, where the older adult may process virtually nothing following the initiation of face-to-face training. In that case, the failure was attributed to the overwhelming of the individual's cognitive resources. Similarly, the precipitation of overwhelming anxiety during the initiation of training could shut down cognitive processing.

The question is what can be done about such inherent anxiety in training situations. Perhaps the most important factor is recognizing that the older learner may be more vulnerable to this state. Thus, if there is group learning related to new technologies or information that older adults may be less familiar with, it is best to *block* training so that younger and older people are not grouped together, and that older adults with less prior knowledge or experience with the subject matter are not grouped with older adults who are more knowledgeable about the topic.

Careful consideration should also be given to the pre-training instructions. The strategy should be to lessen the uncertainty and make the individual as comfortable as possible regarding what will be presented. For example, one could present a brief introduction of the nature of the material, and indicate that even though some of this material may be unfamiliar it will be presented in small chunks and in a very easygoing and nonthreatening manner. Older learners should also be assured that if they do not understand terminology or some concept, that they should not assume it is due to their ignorance or incapability but rather that the training could probably be improved upon, and that they should not hesitate to raise issues or interrupt the trainer. A similar approach could be designed into computer-based training programs such as e-learning programs (Chapter 10). In such learning environments, programs can be made dynamically adaptable to the learner so that if the program detects difficulty on the part of the learner (or the learner signals such a concern), the possibility may exist for a simpler version of the learning modules to be presented.

6.3 Fatigue

Fatigue is a construct related to either mental or physical exhaustion. We speak of *visual fatigue* when one is engaged in activities that place high demands on visual processing. These would include microscope inspection or the need for extended periods of intense vigilance during monitoring in work operations, or long periods of reading, especially under poor illumination conditions. *Mental fatigue* is often associated with extended periods of time in which one is operating at or near one's cognitive capacity, for example, when engaged in long episodes of troubleshooting or problem solving, brainstorming, or learning situations involving a high reliance on attention and memory processes. *Physical fatigue* results from high levels of physical energy expenditure.

Generally, older adults can be assumed to be more vulnerable to each of these types of fatigue. *Homeostatic adjustment capability*, which is, the ability for the body's various organ systems to adapt to varying external demands, not only declines with age, but cognitive processing resources associated with attention, sensory processing, and memory are also subject to reduced capacities with aging, similar to reductions in respiratory and cardiovascular capacities with age.

The most important implication of the potential adverse effects of fatigue concerns the degree to which the instruction should be presented in a massed, as opposed to a spaced, manner (Chapter 4). There are advantages to having massed practice, especially for skills that people need to use repetitively throughout a task's performance (often referred to as recurrent skills), as they strengthen consistent mappings between cues and corresponding actions. In the 4C/ID model of instruction (Chapter 8), where holistic tasks are presented in blocks (i.e., groups of tasks), the reduction of the number of tasks within a block can ultimately negatively affect the systematic process of scaffolding, whereby learners are made to gradually perform more and more of the task on their own. Similarly, it can adversely affect the ability to inject sufficient variability among the tasks within a block for the purpose of enabling the formation of meaningful schemas in the learner's long-term memory. Thus, the solution to possible impending fatigue through the creation of shorter training sessions could backfire in the formation of both recurrent (i.e., procedural) as well as nonrecurrent (e.g., development of mental models and schemas) skills.

Managing this tradeoff represents one of the greatest challenges when training older adults. In the case of a procedure that requires a relatively large number of steps, if fatigue is deemed a problem, then one could break up the training on that procedure or recurrent skill. However, this disruption will need to come at the expense of injecting more review or refresher points into the resumption of training to ensure as much continuity as possible in the highly integrated activity. A similar strategy should be employed if the older learner was performing practice trials associated with a given block of holistic tasks in which a scaffolding strategy was being applied. If the expenditure of effort to complete the block is deemed excessive for the learner, then the block could be divided up, but with some form of mechanism for bridging between the prior task and the impending task that is to be performed, for example, by repeating the prior tasks at the same level of scaffolding, but at a faster pace.

The possibility for older learners having increased physical fatigue at the onset or during training should also be considered. Older people have more chronic illnesses and conditions that can predispose them to fatigue. Our experiences in training older adults have shown that they are more likely than younger adults to raise fatigue as an issue during training. Thus, when training older adults, increased attention needs to be given to factors such as repetitive use of devices such as a mouse (which could produce musculoskeletal fatigue), the effects of glare (which could produce visual fatigue), and seating and other workplace arrangement factors that could produce postures for periods of time that could result in tremendous discomfort for older adults, and thus distract them from receiving instruction.

6.4   Recommendations

- To deal with lower self-efficacy, let older learners know that even younger savvier learners often experience difficulty early in learning.
- Have older adults understand a basic concept or master a basic procedure as a way of instilling confidence in their ability to take on additional instructional materials and provide positive feedback concerning these achievements.
- Convince older learners about the relatively immediate (rather than long-term) benefits or rewards associated with learning the material or task.
- Throughout training, provide encouragement to older learners by informing them that learning new material or tasks will be easier than they may have anticipated, and that as they accrue knowledge and skills subsequent learning will become progressively easier.
- Always inject rest breaks into the instructional session if it appears that the older learner is exerting too much effort. In the case of online

learning environments, use messages to encourage the learner to take breaks and relax before resuming the learning exercises.

- Convince the older learner that following mastery of a task or acquisition of knowledge and skills there will be an opportunity to use these skills to master other learning challenges, and provide them with examples of these possibilities (i.e., make the envelope of potential benefits more obvious to them).
- If the learning can have immediate tangible impacts on the learner's life (e.g., in the enhanced ability to manage her health, commute, or expand her social sphere), point this out during breaks in the instruction.
- Reinforce the idea that engaging in learning may improve intellectual and, depending on the task, physical functioning.
- To diminish anxiety, refrain from using jargon, especially early in learning. Always substitute more user-friendly terms and make use of familiar metaphors to diminish the sense to the learner of confronting uncertainty.
- Consider the application of relaxation techniques that older adults can easily utilize prior to the initiation of reasonably difficult or intellectually demanding learning sessions.
- In group training situations, avoid mixing older adults with younger adults, or with older adults who may be more knowledgeable about the topic.
- Assure the learner prior to the start of instruction that there will not be too much material presented at any one time. Also, assure them that they should not hesitate to interrupt if they are confronting language or terminology that is not understandable, for which instructors should apologize (thus allowing older learners to deflect blame from themselves which might stem from feelings of incompetency).
- Pay careful attention to aspects of the learning environment that could predispose the older learner to fatigue, including repetitive use of input devices, glare from display screens, and poor seating and arrangement of learning materials.

Recommended reading

Brandstädter, J., Wentura, D., and Rothermund, K. (1999). Intentional self-development through adulthood and later life: Tenacious goal pursuit and flexible adjustment of goals. In J. Brandstädter and R.M. Lerner (Eds.), *Action and Self-Development: Theory and Research through the Lifespan*. Thousand Oaks, CA: Sage 373–400.

Cahn, B.R. and Polich, J. (2009). Meditation (Vipassana) and the P3a event related brain potential. *International Journal of Psychophysiology*, 72: 51–60.

Carstensen, L.L. (1992). Social and emotional patterns in adulthood: Support for socioemotional selectivity theory. *Psychology and Aging*, 7, 331–338.

Heckhausen, H. (1997). Developmental regulation across adulthood: Primary and secondary control of age-related challenges. *Developmental Psychology*, 33: 176–187.

Hendricks, J. (2006). Leisure. In R. Schulz (Ed.), *The Encyclopedia of Aging*, 4th ed. New York: Springer, 641–643.

Kanfer, R. (2009). Work and older adults: Motivation and performance. In S.J. Czaja and J. Sharit (Eds.), *Aging and Work: Issues and Implications in a Changing Landscape*. Baltimore, MD: Johns Hopkins University Press, 209–231.

Kanfer, R. and Ackerman, P.L. 2004. Aging, adult development, and work motivation. *Academy of Management Review* 29: 440–458.

Keinan, G., Friedland, N., Kahneman, D., and Roth, D. (1999). The effect of stress on the suppression of erroneous competing responses. *Anxiety, Stress, and Coping*, 12: 455–476.

Lupien, S.J., Gaudreau, S., Tchiteya, B.M., Maheu, F., Sharma, S., Nair, N.P., Hauger, R.L., McEwen, B.S., and Meaney, M J. (2006). Stress-induced declarative memory impairment in healthy elderly subjects: Relationship to cortisol reactivity. *Journal of Clinical Endocrinology and Metabolism*, 82: 2070–2075.

Maehr, M.L. and Kleiber, D.A. (1981). The graying of achievement motivation. *American Psychologist*, 36: 787–793.

Sharit, J. (2012). Human error and human reliability analysis. In G. Salvendy (Ed.), *Handbook of Human Factors and Ergonomics*, 4th ed. New York: J. Wiley & Sons, 734–800.

chapter seven

The human information processing system—The "learning engine"

7.1 Revisiting age-related declines in cognitive abilities

Why should issues surrounding training and instruction be any different for older adults than for other population subgroups? One possible reason is that older adults may have much less motivation to engage in new learning, thereby implying the need for different instructional strategies. As discussed in Chapters 4 and 6, motivation can be a powerful mediator of learning, and with older adults some unique factors may need to be considered that could influence their inclination to take on new learning challenges. Overall, however, there is not sufficient evidence to warrant tailoring instructional programs to older adults based on their possibly distinctive motivational profiles.

Assuming a sufficient degree of motivation exists for learning, a key consideration in designing training and instructional programs, particularly for older adults, concerns the learner's information-processing capability. Learning typically involves the need for the elucidation of new information in the form of facts, ideas, and procedures. Although possessing lifelong knowledge or experience that is positively associated with this new material can make it less taxing for older adults to absorb it, there are a number of structures or mechanisms comprising the human's information-processing system, which we sometimes refer to as *processing resources*, that are likely to show some degree of age-related decline. These processing resources, which are associated with processing or transforming information within the human information-processing system, include perceptual processing, storage and manipulation of information in working memory, and transfer of information to and from long-term memory.

When documenting age-related declines in such abilities, two broad categories of cognitive ability that are often recognized are *fluid ability* and *crystallized ability* (Chapters 2 and 4). Fluid abilities broadly reflect abilities that are involved in new learning or problem-solving performance. They generally peak somewhere in the 20s or 30s and then gradually decline with increasing age. In contrast, what is referred to as crystallized intelligence remains relatively stable or increases throughout the life span at least until about age 70.

This distinction is relevant to aging and learning, as it suggests continuous declines with age in fluid cognitive activities such as detecting and inferring relationships between variables, novel problem solving, memory of unrelated information, transforming or manipulating unfamiliar information, and real-time processing in continuously changing situations. It also implies increases or stability in crystallized cognition through a good part of the adult life span as reflected in measures of acquired knowledge.

These age differences in cognitive abilities, however, are typically inferred from standardized testing, which may underestimate the potential capabilities of older individuals. Increases in cognitive ability into very late adulthood may be found if more comprehensive and relevant assessments of knowledge were available. Also, there is a great deal of evidence in the literature that indicates older people are able to offset declines in cognitive abilities with knowledge and skills acquired through experience.

Another problem with such broad characterizations of age-related patterns in cognition is the large variability in performance across older people. The implication of such variability is that on some tasks older people perform at the same level or even outperform much younger adults. This was, in fact, found in a number of studies involving computer-based information search tasks (e.g., Czaja et al., 2001) where, despite the expected trend of performance decline with increasing age, there was considerable variability in performance. In these studies, a number of older adults actually outperformed their much younger counterparts, which should not be that surprising in light of the variability in fluid cognitive abilities that is found across age as depicted in Figure 2.6 in Chapter 2. An important implication from these types of studies is that, depending on the nature of the task, certain cognitive abilities may provide older adults with the capability for compensation for declines in other abilities.

Training and instruction involve imparting information. Consequently, a number of fluid abilities will be needed to capture, retain, and perform the necessary manipulations of this information. To understand why this may be problematic for many older adults requires an examination of some of the features and mechanisms of human information processing and the consideration of how these mechanisms may be affected by aging.

7.2   An overview of human information processing

A basic model of human information processing from a human factors perspective is presented in Figure 7.1. We present an enhanced version of this model that more directly relates to training and instruction (including

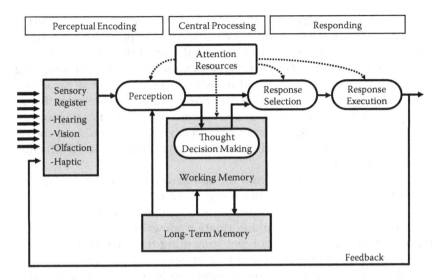

Figure 7.1 An information-processing model. (Adapted from Wickens et al., An Introduction to Human Factors Engineering, 2nd ed., New York: Prentice-Hall, 2004.)

multimedia learning) in Section 7.5. The fundamental aspect of the model in Figure 7.1 is that it portrays human information processing as a flow of information between various information stores that humans possess and consisting of various transformational processes that can be applied to this information.

In this model, *attention resources* can be thought of as a pool or supply of mental effort or energy that people possess. This supply not only has a natural upper limit but it can also vary over time, as any learner can attest to during times of mental exhaustion or physical fatigue when it becomes difficult to focus one's attention on incoming information. The dashed arrows emanating from attention resources indicate that attention can be allocated to various cognitive processes as required. For example, consider the sensory information received by our body's various receptor cells. This input information results in sensory traces stored in sensory registers that collectively comprise *sensory memory*. Resources of attention can then be used to select sensory channels for further processing, as indicated by the dashed arrow to the perception box (this process may often initially entail a path through working memory, which then instigates the specific information to select; however, this is not directly observable in this model). This process is known as *selective attention*.

There are a number of factors that can influence the selective attention process. Some of these factors influence attention by working from the *bottom up*. That is, which channels of information the person attends to and, conversely, which are filtered or ignored, are affected by external features associated with stimuli or events. An example of such a bottom-up factor is the *salience* of the event or signal. Salient stimuli will generally capture our attention, a fact that can be exploited in training or designing instructional material for older adults with declining cognitive capacities. For example, in a computer-based administration of practice problems (Chapter 10), the ability to locate critical buttons or links to ensure continuity in interacting with the training system's interface would depend on making these stimuli salient, but not to the point that they are distracting when not needed.

Top-down factors, in contrast, are not driven by environmental cues but rather by one's internal knowledge about the situation. Examples of two such factors are *expectancy* and *value*. Expectancy is a powerful force. If people expect to find useful information somewhere (e.g., in a particular menu), or expect to hear or feel something that may be meaningful (pressing on that button will lead to a sound indicating the device took a reading), they orient attention by sampling information in a direction consistent with those expectancies. Likewise, if a person has knowledge about how beneficial some channel or source of information is (e.g., the home page always shows the directory needed for starting a search all over again), or how costly it might be to miss that information (not looking at the upper left-hand side of the screen may lead to missing the answer), attention is more likely to be directed toward those channels.

Because there is a tendency for people to minimize cognitive effort when performing various cognitively demanding tasks, the selective attention process may be inhibited if it requires too much effort. This is especially true for older adults who, to begin with, may have reduced attention capacity. Thus, if an instructional program requires the older learner to navigate through a relatively large number of screens or perform a long series of steps to obtain needed information, despite a reasonable expectancy that this information will be there and will be of value, the required or perceived effort may inhibit such actions from being performed.

During training and instruction the sensory channels to which attention is allocated are primarily the visual and auditory senses, but they can also include the haptic senses if materials or devices to be handled are part of the instructional experience. Attention can also support other aspects of performance, as indicated by the other dashed arrows in Figure 7.1, through its distribution among other processes. This process represents *divided attention*, which is often necessary during instruction. However, forcing attention to be divided unnecessarily is one way of imposing an impediment to learning, and one that could adversely affect older individuals to a greater degree than younger adults.

Figure 7.1 also shows that actions can generate new information to be sensed and perceived (the feedback loop). This is an especially important consideration in many of today's training programs, whereby learning is by doing rather than by attempting to absorb large amounts of factual information passively (Chapter 8). With older adults, it implies the need for sufficient attention not only to process and understand what was just done, but also what happened as a result of these actions. This is another area where instructional programs could benefit from interventions targeting older learners. Specifically, designers of instructional programs need to ensure that limited resources of attention are not overloaded when the learner simultaneously attempts to process information related to actions and the results of those actions. Finally, it is important to note that this overall sequence of information processing as depicted in Figure 7.1 can start anywhere, as is emphasized when some of these information-processing mechanisms are examined in more detail.

What typically happens when information receives the attention of the learner? Referring to Figure 7.1, this information is processed further in the perception stage, where meaning from sensory information, such as printed text, icons, graphs, pictures, and spoken words, is extracted. Typically, meaning is obtained through comparison of that information to that which the person has in long-term memory (LTM). The process of adding meaning to information can also occur without the need for attention if the information is very familiar to the person.

Following perceptual processing, the person may immediately react with a response. This often occurs with *skilled-based tasks* in which cues have become "hard-wired," through consistent mapping, to particular types of responses. In some instructional situations, it is essential that the learner rapidly and accurately associate certain cues with certain responses, as these cues or signals may appear frequently. Thus, the ability to process this information with a minimal investment of attention resources provides extra capacity for other aspects of the tasks. However, what is more likely to occur during the initial stages of training or learning (Chapter 4) is that the information will require further processing in working memory.

Working memory (WM) refers to the short-term store of whatever information is currently active in the information-processing system. As noted, it is a kind of workbench of consciousness in which people visualize, plan, compare, evaluate, transform cognitive representations, understand, make decisions, and solve problems. Unless this information is rehearsed, which allows it to be encoded in LTM, it can decay rapidly. As described below, many of the concerns associated with providing instruction to older learners are associated with WM, in view of its fragile nature and capacity constraints.

Finally, depending on the situation, some type of response, either manual or voice, may be required. The selection of these responses may, as discussed earlier, be directly linked to perception (i.e., they have become hard-wired), or they may be determined after decision making or problem solving based on WM and LTM processing.

7.3   A closer look at the role of information processing

The overview of the information-processing model given above was intended to highlight the principal mechanisms involved in perceiving and processing information and their potential implications for learning, especially for older adults. In this section we provide more details concerning these processes. We begin with perception, where meaning may be added to sensed information through comparison of that information with information stored in LTM.

7.3.1   Perception

Perception is a process that must be capable of transforming a host of incoming sensory information into a mental representation, code, or some entity that can be understood. This is a critical process in instruction because these representations often become the concepts that will undergo further elaboration during learning. During the earlier stages of learning, it is not likely that the presented information will require such little investment in attention that it could trigger an immediate response (Figure 7.1), although this might be an eventual goal of training. More realistically, during a good part of training it is expected that there will be a dependence on LTM for clarifying the information being presented. For recurrent activities—that is, activities or procedures that often need to be repeated within the larger context of more cognitively challenging activities—automatic processing of sensory information can usually be achieved if there is consistent mapping of the stimuli to responses.

Three processes, which can occur concurrently underlie perception: bottom-up feature analysis, unitization, and top-down processing (Wickens et al., 2004). *Bottom-up feature analysis* represents effortful perceptual processing that typically occurs when unfamiliar events or stimuli are encountered. Handling such input entails that the stimulus pattern be decomposed into more elemental features (hence the name *feature analysis*) with the burden on the learner to use these more elemental components to establish a meaning. With bottom-up feature analysis, the meaning of the perception will usually be retrieved from LTM relatively slowly and with effort.

This type of perceptual processing has a number of implications for the older learner. One problem is that despite the potential for a reasonable wealth of knowledge stored in LTM, if the domain of instructed knowledge is relatively unfamiliar to the older adult, there may be little in the way of stored information that could be used. Another issue is that with aging comes a general slowing in all information-processing activities (Chapters 2 and 4). An effortful process that relies on decomposition thus may make it difficult to retrieve information from LTM. Retrieval would be especially hard if the pieces of knowledge that may be of value during instruction are not adequately associated with other relevant information or have not been encountered frequently, as is elaborated on in the discussion on LTM. A third issue is that any factor capable of degrading the stimulus characteristics (e.g., poor illumination, masking noise) can interfere with bottom-up processing, and these are factors to which older adults are typically more susceptible. The basic idea behind solutions to these issues is to make readily available those critical features that are needed to attach meaning to the perception, and to do so at a low investment of attention for the learner.

The second process of perception, *unitization*, represents an automatic processing of sets of features that are familiar, presumably because they have occurred together very frequently (e.g., the processing of the letters in the word "the"). The combined representation of these sets of features is stored in LTM. The third process governing perception, *top-down processing* can be thought of as the ability to correctly guess what a stimulus or event is, even in the absence of clear physical (i.e., bottom-up) features necessary to precisely identify it. These guesses are based on expectations that, in turn, are based on past experiences stored in LTM.

If older learners are suspected of having inadequate information in LTM to perceive learning materials correctly, especially early in training, a good strategy is to avoid, as much as possible, having such learners resort to bottom-up feature analysis. Not only is such processing likely to fail, but it requires that all other environmental conditions (such as illumination) be ideal. Generally, bottom-up feature analysis, by virtue of being effortful, exacts resources that for older adults in particular should be reserved for other information-processing activities.

A task analysis (Chapter 10) applied to the instructional materials can help identify all or many of the fundamental stimuli that comprise the training program (e.g., directional arrow configurations or system message icons) whose interpretation may be critical for carrying out actions or making decisions. These items could then be given appropriate consideration within the instructional method that is implemented (Chapter 8). The task analysis should also identify where sets of symbols or cues logically or often go together, to promote unitization, which reduces the cognitive demands of perception. Another important consideration to ensure that stimuli are perceived correctly and without large investments of effort is to point out whether certain stimuli, such as icons, pointers, or screen configurations, appear in certain contexts or modes of device operation. This type of emphasis can lead to more effective top-down processing, whereby the older user now has a better ability to guess what some input pattern may signify.

Many training or learning situations rely on the use of some form of display. For example, a display on a computer screen may be used in a simulation to illustrate various states or conditions that the learner must recognize in order to assess a situation or to respond appropriately. The prevalence of the use of displays in task training implies the need to consider display design principles and issues related to information visualization that have the potential to greatly impact learning and task performance for older adults. Principles of display design are generally based on ensuring that the information contained in the display is compatible with human capabilities associated with perception, attention, and memory.

One such principle, the *proximity compatibility principle* (PCP), suggests that decision performance using a particular display will be best when the information contained within the display matches the demands of the task. For example, if a task requires divided attention in order to combine, compare, or otherwise integrate the individual pieces of task information into a holistic judgment, such a task would be said to have *high task proximity*. When multiple pieces of information need to be considered independently or when focused attention is needed to extract a single value from the multiple pieces of information, the task would be considered to have *low task proximity*. Similarly, *display proximity* refers to the level to which the perceptual objects in a display are integrated. When the display's information elements represent uniquely different pieces of information, the display is said to be a *low-proximity display*. In a *high-proximity display*, the separate perceptual elements represent multiple pieces of information related to the same object.

The implication of the PCP is that task and display proximity should be matched as much as possible: high-proximity tasks are best supported by high-proximity displays and low-proximity tasks are best supported by low-proximity displays. The premise is that mismatches in task–display proximity are likely to have more severe consequences on task performance for older learners due to the additional demands on attention that would be needed to resolve the mismatch. Thus, if an older worker were being trained to find the value of a particular variable on a display as a basis for making a decision, a low-proximity display would decrease demands on focused attention and thus would make it easier for that person to identify the signal within the potentially "noisy" display. However, suppose the task entailed an intervention when the combined values of multiple variables exceed a threshold value. Then a high-proximity display, for example, one that takes on a particular geometric shape that corresponds to the values of the multiple variables and thus can highlight, perceptually, the collective status of these variables, would be appropriate.

There are several other display design principles that should be given particular attention when older learners are considered. The obvious ones involve ensuring legibility (or audibility in the case of auditory signals), and that the need for making absolute judgments be minimized and kept to a low number of levels if such judgments are necessary. *Legibility*, which relates to visual issues such as contrast, illumination, and visual angle, and auditory issues such as noise and masking, is a necessary condition for ensuring usable displays for older users, and is critical for making sure that the benefits deriving from the implementation of other principles, such as the PCP, can be obtained. However, it is far from a sufficient condition for creating usable displays. Absolute judgment tasks refer to those tasks that require the human to judge, often through interaction with some type of display, the level of a represented variable on the basis of a single sensory variable such as color, size, or loudness. Although the number of levels specified for tasks requiring these types of judgments is generally between five and seven, these are difficult kinds of tasks and if possible should be avoided altogether with older trainees or restricted to even lower levels, such as two to three.

Another display design principle with important implications for older users concerns *information access cost*. Often, when a user chooses information from displayed information sources, a certain amount of selective attention effort is needed. The access cost refers to the cost in the time or effort that is needed to move selective attention from one display location to another. When creating displays directed at instruction or performance by older adults, the imposition of such costs could undermine learning or performance. This net cost can be minimized by, for example, ensuring that different pieces of information that are likely to be used in sequence be placed in such a location that the cost for the learner of traveling between them is small.

During design of instructional programs that involve displays, one question that frequently arises concerns how much information to place in the display and how much information the trainer should assume the learner will be able to maintain in memory. Information that the person is required to retain in memory is often referred to as *knowledge in the head*. Display design guidelines influenced by principles of memory often dictate that such memory be replaced with displayed visual information, referred to as *knowledge in the world*.

Older learners in particular should not be required to retain important information solely in WM or have to retrieve it from LTM. However, too much information in the world can create problems with requirements for legibility and selective, focused, and divided attention. Good instructional display designs may require anticipating the kinds of information that are likely to be retained with familiarity and alter, accordingly, the content of displays as training progresses, with the option of bringing back the information removed if the instructional process suggests the need to do so.

7.3.2   Working memory

WM is the short-term memory or temporary store that keeps information active while it is being used or until it is needed. Typically, processing information in WM requires that information has undergone some degree of perceptual processing, or that it has been residing in LTM from where it must be brought into WM (as would occur when one initiates a thought process). In many cases, factors related to WM are likely to affect the success of training and instruction for older people, as WM has the potential to serve as a bottleneck in absorbing and understanding needed information.

Consider the following anecdotal episode. One of the experimenters affiliated with our research center was in the process of training an older woman on the use of a computer-based system when the participant interrupted her and said: "I know you are trying to tell me all these things that I need to know and trying to help me but all I'm hearing is blah, blah, blah." We refer to this phenomenon, which corresponds to an overload in WM, as the *blah blah blah effect*. This effect can be influenced by a number of factors such as:

- The degree to which there is a gap or mismatch between the person's prior knowledge (i.e., knowledge stored and retrievable from LTM) and the knowledge being introduced
- The rate at which information is being introduced
- The sequence in which the information is introduced
- The opportunity for the learner to pause and apply aspects of the information being introduced
- The conduciveness of ergonomic or environmental conditions to taking in new information
- The physiological state of arousal of the learner

In fact, in one study the largest barrier older adults perceived with regard to their learning needs concerned cognitive matters (Purdie and Boulton-Lewis, 2003). The types of problems that were mentioned included not being able to absorb things well, not being able to hold many things in mind, not remembering sequential procedures such as how to start a computer, and not being able to concentrate for too long. Although the latter issue relates to attention capacity, for the most part these perceptions reflect diminished WM capacity.

Whether attention is directed inwardly (e.g., during thoughtful introspection about a topic or when planning what to do) or, as often occurs in training, outwardly toward selecting information from various training sources for further processing, there is only a limited amount of information that can be brought into WM. This *capacity constraint* is even more critical with older learners as cognitive capacity declines with aging (Chapter 2). Trainers thus must be very careful about how much information is being presented to the older learner. Even if the amount of information appears reasonable, generalized slowing in information processing (Chapters 2 and 4) requires that the trainer also be aware of the rate at which information is being presented.

In addition to capacity, the other critical WM constraint is time, which concerns how long the information can be kept active. The strength of information in WM decays over time unless it is periodically reactivated, a process that is sometimes referred to as *maintenance rehearsal*. If one were to assume that with aging this process of decay is accelerated, that is, the strength of the traces weaken more rapidly, the implication is that more effort would be needed for maintenance rehearsal. Such rehearsal would help ensure that material has been adequately encoded so that material that was to follow, and which was dependent on the prior learning material, would have a better chance of being understood. In many learning situations, instructional design may afford for both the time and the capacity constraints of WM to be manifest jointly. For example, consider the case where a learner must read through material that consists of a fairly lengthy string of items. By the time the end of the string has been reached, the earlier information in the string will have already likely decayed, especially for older adults who, as has been suggested, may have WM systems governed by more rapid decay rates.

One way to optimize the process of maintenance rehearsal is to partition the presentation of information into batches comprised of one or more *chunks* of information that are within the person's WM capacity limits, and to ensure that sufficient time is provided between these batches of information for encoding. A chunk is essentially a unit of WM space, and is defined jointly by the physical and cognitive properties that bind items together within the chunk. An icon with an associated text label could represent a chunk, as might the information associated with three things you need to check before starting a device. In the latter case, if the three entities were random facts that had no relationship to one another, then these entities, at least initially, would correspond to three chunks. However, if the facts were all associated with one another (e.g., they represent conditions to ensure a device is ready for use) and the learner has, through maintenance rehearsal, adequately encoded them as a set, then upon retrieval from LTM they would occupy one chunk of space. Similarly, a graphical depiction of a status indicator containing a particular geometric configuration and corresponding message or a meaningful sequence of several alphanumeric characters can define chunks that are based on cognitive binding, which is analogous in some respects to the process of unitization in perception.

There are several advantages to identifying material and teaching that material to the learner in the form of logical chunks, for example, by pointing out to the learner that a set of information sources should be processed (i.e., selected through attention) as a single entity. First, chunking allows for easier rehearsal of material, increasing the likelihood of transfer to LTM. Second, encoding information in LTM in the form of chunks that constitute meaningful associations aids in the retention of the information in LTM, making its transfer to WM more reliable (Chapter 5). Third, when this information needs to be brought back into WM from LTM, chunking increases the capacity of WM storage. This would enable the learner to hold more information in WM during the learning process, which increases the prospects of successful training.

Although there is a clear advantage to chunking with respect to WM storage capacity, increases in the size of the chunks or changes in their features can result in WM capacity being exceeded, which makes it very difficult to pin down the capacity of WM to a number of chunks. The traditional upper limit of 7 ± 2 chunks, which has often been associated with WM capacity, thus needs to account for the various attributes that can characterize the chunks. Seven chunks representing the random digits associated with a telephone number may possibly be within WM capacity limits, but seven chunks, each corresponding to a highly complex rule comprised of many facts, may not be. With more recent estimations suggesting that WM capacity may be closer to 4 ± 1 chunks, and even less in older adults, there is an even stronger argument for exercising caution with regard to the amount of information that is presented to these learners during instruction, in addition to the manner in which this information should be presented.

7.3.3   A model of working memory

A three-component model of WM proposed by Baddeley (1986) is fundamental for providing a number of insights for training learners in general, and especially those who are older. The three components in this model are the following:

- *The phonological loop*: This storage system represents *verbal information* in an acoustical form. This verbal information can derive from the visual modality (e.g., printed text) or the auditory modality (e.g., speech). This information is kept active or rehearsed by articulating sounds or visually presented words, either vocally or subvocally. Presumably, auditory spoken information gains *automatic and obligatory access* to the phonological store, so that the phonological memory trace can be disrupted by unattended spoken material. This can be an important consideration with older learners as aging has been associated with a reduced ability to suppress distracting information.

- *The visuospatial sketchpad*: This storage system holds visuospatial information in an analog spatial form (e.g., visual imagery such as pictures or stimuli, including verbal stimuli that are pattern-based and consequently spatial in nature), and thus helps ensure the formation and manipulation of mental images. Presumably, this storage system can be disrupted by irrelevant visually presented items such as pictures or patterns, or by concurrent spatial processing (even if it is through mental imagery).
- *The central executive*: This acts as an attention control system that coordinates information from these two *memory storage subsystems*. It selects certain streams of incoming information and rejects others, and selects and manipulates information in LTM. Generally, it is responsible for thought processes leading to decision making and problem solving.

The ability for people to integrate verbal and spatial information effectively may depend on individual differences in spatial ability. There are a number of tests available for measuring an individual's spatial ability, and several studies (e.g., Gyselinck et al., 2000) have found that highly spatial people benefited more from the concurrent presentation of verbal (e.g., text) and visuospatial (e.g., visual animation) information as compared to when each of these types of information was presented successively. Likewise, some individuals may have higher phonological memory capacity (e.g., as measured by the digit span test), and may prefer a text-only format to an illustrations-only format. These tendencies have important implications for instruction as they imply, for example, that some individuals may be more or less suited for presentations of learning materials such as multimedia that integrate pictures and text. For older learners who may already be at a cognitive disadvantage, it suggests that instruction could be made more effective if it were based on the learner's visuospatial or phonological capacities. The idea of customizing instructional materials and presentation modes to better conform to these capacities, which may be able to be estimated through simple tests, represents an interesting possibility in the area of training directed at older adults.

Theorists are not always in agreement about exactly what is going on in WM. However, the distinction between visual–spatial and verbal–phonetic types of information is crucial, and is an important aspect of *multiple resources theory* (Wickens et al., 2004). According to this theory, there isn't one single undifferentiated pool of attention, so to speak, from which attention can be allocated but rather multiple relatively distinct pools that can be characterized into four dichotomous dimensions. Each of these dichotomies has two different levels, with one of the two levels of each dichotomy using different *resources of attention* than those used by the other level. To understand what this means, consider a person performing a job that consists of two tasks that need to be time-shared, such as watching a display for signals while processing some other set of information. To the extent that these two tasks demand common levels on one or more dimensions, time-sharing is likely to be worse, and one or the other task will decrease farther from its single task-performance level. The four dichotomous dimensions of the information-processing system are

1. *Stages*: Perceptual/cognitive (*early*) processes versus response (*late*) processes (Figure 7.1). Thus, resources of attention used for perceptual and central processing (such as searching or listening) are largely separate from resources used for response selection or execution.
2. *Modalities*: Auditory versus visual. The difficulty associated with performing two tasks due to the need to process two visual or two auditory channels can be lessened if the input from the two tasks is made cross-modal (e.g., studying the configuration of a machine through use of the visual channel and learning about how certain components of the machine work through an accompanying voiceover).
3. *Codes*: Spatial versus verbal (print/speech) in the early (perceptual/cognitive) processing stage and manual versus vocal in the later responding stage. Thus, there should be a greater likelihood of successfully processing pictures and voice during instruction, as this strategy not only uses different codes but also different modalities.
4. *Visual channels*: Focal versus ambient. The visual resources used for reading and interpreting symbols are different from those used to, for example, maintain balance.

The dimension related to codes is most relevant to the model of WM. Essentially, visual–spatial and verbal–phonetic types of information represent different WM codes, and each of these codes seems to process information somewhat independently, using its own resources. It would seem, especially for older learners, that distributing information across codes makes more sense than more extensive use of one code, as there would be less chance of overloading the limited cognitive or storage capacity associated with the processing of an individual code.

However, there is also the central executive component of WM to consider, which has the responsibility for integrating the information from these two storage systems. But what is it that is being integrated? Presumably, verbal information, whether it is derived from the auditory (spoken) or visual (written text) channels would first need to become

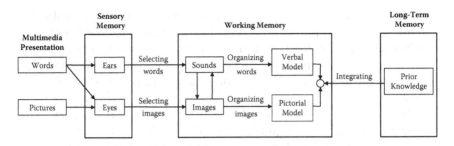

Figure 7.2 A cognitive theory of multimedia learning. (From Mayer and Moreno, *Educational Psychologist*, 38: 43–52, 2003. With permission.).

organized into a verbal model, and the analogical–spatial information would need to become organized into a pictorial model (Figure 7.2). The selection of information from sensory input, whether it is from the visual or auditory sensory channel is mediated through the perception process but, as discussed later in the chapter, is very much influenced by thought processes in WM, of which LTM can play an important role. For example, the learner may decide to disregard some aspects of the auditory (e.g., narrated) input based on prior knowledge in LTM that suggests this information is redundant with already learned or presented information and is not worth allocating attention to. In any case, once information is brought into WM, its two storage systems must generate the corresponding verbal and pictorial models (Figure 7.2). Too much verbal or image-based information can overload the cognitive processing capacities of any of these storage systems and thereby undermine the development of these models in WM.

Even if the amount of incoming information that requires processing in each of these storage systems is within capacity limits, if extensive retrieval of LTM information is needed to organize these models, then the introduction of this additional information from LTM into WM can overload these individual storage systems. The greater the amount of prior knowledge the individual has in LTM, and the more accessible this knowledge is (e.g., because of its frequent or recent use, or because it is well-organized in LTM), the less would be the load within WM upon its retrieval for the purpose of generating verbal and pictorial models. This is why it is so critical to ensure that older learners receive the proper presentation sequence of learning materials during instruction (Chapter 8) and to provide these learners with sufficient time to rehearse this material so that it becomes adequately stored in LTM.

The process of actually integrating or coordinating this verbal and spatial information in WM, which is presumably the role of the central executive, can also benefit from knowledge residing in LTM. What is less clear is the toll on cognitive resources of attention that this central executive component of WM might exact. If these demands are substantial, and if they are particularly sensitive to aging, then the benefits of distributing input instructional information across the verbal and spatial–analogical codes may become negated by the effort the learner may need to invest in integrating the information from the two WM storage systems. In fact, this remains a dilemma in assessing the benefits of multimedia for training older adults (Chapter 10): can the benefits associated with enhancing limited cognitive capacities in WM storage systems be offset by the costs associated with having to integrate information from these memory systems?

7.3.4   Long-term memory

In the preceding discussion of human information processing within the context of instructional design and older adult learning, the emphasis was predominantly directed toward WM. This is justifiable, as WM is the primary bottleneck in the learner's ability to select, integrate, and ultimately make sense of presentation materials. Although the complementary role of LTM (Figure 7.1) was repeatedly alluded to, there was no explicit reference to its nature. By no means was this omission meant to minimize the influence of LTM on older adult learning.

The structure and mechanisms associated with LTM constitute a highly complex topic. In keeping this topic manageable and relevant to the theme of this chapter, LTM is considered simply as a mechanism for storing information that may be required at a later time. This mechanism may be activated at various times: to initiate a thought process concerning how a task should be performed for a planned work procedure, when confronted with signals from the environment such as readings on a display screen that need to be interpreted, or when recalling material that was just taught in order to comprehend learning material that will follow.

Irrespective of how we may use the stored information, what is important to recognize is that the "pieces" of information contained in LTM result in the formation of multitudes of associations. Indeed, LTM can be viewed as a parallel distributed architecture that is continuously being reconfigured depending on incoming information, whether through the sensory apparatus or through WM activation by virtue of thought processes (Figure 7.1). When different pieces of information are processed (which can be text, sounds, images) at the same time in WM, as typically occurs during instruction or learning, these items of information become associated in memory. Pieces of information that are meaningful when considered together can then form the basis for later reactivation from LTM. What this means is that activation of parts of associations, for example, the conditional part of an IF–THEN rule, will increase the likelihood of other associated information becoming activated as well. This can lead to increased efficiency in learning and skill acquisition (Chapter 4).

One way in which learning and skill acquisition become more efficient with continued instruction and practice is when the meaningful associations between the items stored in LTM result in the formation of chunks of information. As previously discussed, having information available in such packet form can reduce the load on WM during retrieval of information from LTM during the course of instruction. These meaningful or related associations in LTM comprise what are often referred to as *associative networks*. Because much of the knowledge obtained from learning is *semantic knowledge* (i.e., the basic meaning of things), the types of associative networks that represent how knowledge is largely stored in LTM can be considered to be made up of *semantic networks*.

Information in LTM often tends to be organized around topics or concepts. The entire knowledge structure about a particular topic is often termed a *schema*. Schemas associated with how a product or system works are often referred to as *mental models*. When one has a mental model about something, it allows for a set of *expectancies* about how that entity will behave (e.g., if temperature of this process is increased, the pressure will increase proportionally). The organization of these schemas and mental models in LTM are also in the form of associative or semantic networks.

Early in the course of learning and skill acquisition, when concepts are still not well understood and meaningful associations are still not well formed, these schemas and mental models may not yet exist or may be incomplete or incorrect (Chapter 4). Being able to impart useful schemas and mental models to learners, especially older learners, can have a powerful effect on learning. Specifically, it allows for prediction, which is normally a very effortful cognitive activity, of what would occur in various circumstances, and for generalization to new situations.

During instruction, learners are faced with having to recall previously taught material (and possibly previous material they may have learned in other situations) in order to support the current instructional topic. Often, this requires retrieving information from LTM, such as an isolated fact or a rule or concept that may comprise an associative network of items. The availability of information from LTM so that it can be used by WM mechanisms to support ongoing instruction is influenced by two major factors: the strength of the pattern itself (often referred to as *item strength*), and the strength of the *association* of that information with other items in memory.

The more frequently information of a particular type has been activated (e.g., the rehearsal of a rule), the stronger is the memory trace, and thus the easier it is to reactivate that trace when needed in the future. Also, the more recently it has been activated, the easier will be its retrieval. These effects were referred to earlier as *frequency* and *recency* effects. When the associative strength is weak, this means that when we try to activate or retrieve a "target" item of information by first activating something with which the item has a semantic association (which is often done by thinking about several semantically related concepts), the activation does not sufficiently spread to the target associative concepts. By not having enough activation at the right place, recall of the desired information is not triggered, which is a phenomenon many of us frequently encounter and a source of frustration, especially during learning situations.

Generally, the failure to retrieve information (or to retrieve correct information) from LTM will occur due to decays of item strength or association strength (Chapter 5). More specifically, these decays occur as a result of weak strength due to low frequency or not being recent, weak or few associations with other information, or interfering associations (e.g., due to many associations needing to be acquired in a short period of time). Thus, to increase the likelihood that information will be remembered at later points during the course of instruction, the information should be processed frequently in WM (to increase item strength) and together with other information in a meaningful way (to increase the strength and number of associations).

As with WM, there is also evidence of age-related declines in associative (long-term) memory (Naveh-Benjamin, 2000). The potential exists, however, for much of the vulnerability deriving from WM limitations to be offset by strategies that strengthen the content and organization of information in LTM. Carefully thinking out the kinds of information that could play a supportive (or priming) role in the learning situation or the sequence with which information is presented are examples of strategies that could be used to minimize the constraints imposed by WM (Chapter 8). In addition, with older adults it is necessary to pay careful attention to instructional protocols governing distributed practice, refresher and priming effects to bridge the gaps between learning sessions, repeated integration of previously learned material into current topics, and other factors discussed in Chapters 4 and 5 to ensure that meaningful information is strongly and reliably secured in LTM. As a result, this information can be used, in conjunction with WM, to enable effective ongoing learning.

7.4   Cognitive load theory and working memory

Theories or perspectives related to human information processing have had a major influence on the field of instruction. Particularly well known is Sweller's (1994) *cognitive load theory* (CLT), which emphasizes how limitations in human information processing, and particularly in WM processes, could account for problems learners encounter during instruction. In so doing, CLT presumably can provide insights into the design of instructional procedures that could facilitate learning by avoiding such constraints. WM and LTM are accorded critical roles in cognitive load theory.

One key to training and instruction then is to identify cognitive load constraints that can obstruct the learning process. In Chapter 5, three types of cognitive loads associated with CLT were discussed: *essential* or *intrinsic cognitive load, extraneous cognitive load*, and *generative* or *germane cognitive load*. Generative cognitive load is a type of cognitive load that can derive from the deeper forms of cognitive processing undertaken by the learner to accentuate the learning experience, for example, through well-designed worked examples and practice exercises. Managing this load from an instructional design perspective is mostly about making use of good instructional practices (e.g., see Chapter 8), for instance, those that enable learners to build into their LTM the requisite knowledge and skills prior to moving forward in the learning program in order to facilitate deeper learning.

Here, the focus is on the intrinsic and extraneous load constraints. Intrinsic cognitive loads derive from the nature of the material to be learned, and in particular, the degree of interactivity associated with the elements comprising the task. The underlying premise is that the ability to learn elements in isolation imposes less cognitive load than learning information simultaneously. This makes sense, as the requirements for integration of task elements will be subject to a greater degree to both the capacity and time constraints of WM. Interactivity can also impose a significant cognitive load even when the number of elements is relatively small. In the case where task elements can be learned in isolation without reference to other elements, task performance can still become undermined by linearly increasing the number of such elements.

Extraneous cognitive loads refer to the way information is structured, and thus include those aspects related to the instructional format of the task. Often, these loads are imposed by engaging learners in cognitive activities that are irrelevant to the learning process, and can be brought about through various techniques and procedures associated with the presentation of instructional material. A classic example, as implied in the earlier discussion on WM constraints, is when the learner is forced, by virtue of the design of the instructional process, to divide attention between multiple sources of information and then to expend effort to hold these sources in memory in order that they be mentally integrated. Generally, these situations involve sources of information that, in isolation, may have little meaning, but in combination are essential to the learning process. However, such extraneous cognitive load can also arise when the multiple sources of information being presented are each intelligible in isolation. For example, additional and unnecessary cognitive processing can be instigated when multiple sources of information are redundant in the information that they convey. As discussed below and in Chapter 10 on multimedia instruction, although redundancy is generally a positive design feature, if designed poorly it can add extraneous load, for example, when the learner attempts to read text statements to confirm verbal information being presented through narration.

Extraneous load can also arise from the use of other techniques, such as the presentation of "seductive" materials to generate more interest within the learner. To the extent that this material does not promote the development of rules, schemas, or automation of cognitive processing, such materials are extraneous and thus potentially capable of disarming limited resources of attention from the learner, and especially the older learner. If motivation is a concern with older people in training or instructional situations, other approaches should be used that do not constitute extraneous cognitive loads. Similarly, the presentation of background music, which a designer may consider as a means for relaxing the older learner, could also ultimately impose extraneous load.

The limited capacity of WM and its implications for cognitive loading were pursued more extensively by Mayer and Moreno (2003) and Moreno (2006), primarily within the context of how learners integrate visual and auditory material within multimedia instructional environments. In Figure 7.2, there is a Sounds WM subsystem for processing auditory input and an Images WM subsystem for processing visual input in the form of text and pictures. Strictly speaking, these two subsystems do not correspond to the phonological loop and visuospatial sketchpad subsystems in Baddeley's WM model (Section 7.3.3). For example, in Baddeley's (1986) model verbal information presented visually is presumed to be immediately processed by the phonological loop subsystem.

Nonetheless, these two WM conceptualizations can be more or less reconciled because in the model depicted in Figure 7.2, verbal information in the form of visual images (i.e., visually displayed text) can be processed by the Sounds component of WM (as indicated by the upward arrow from the "Images" box to the "Sounds" box) in order to generate the verbal model. In effect, the conceptualization of WM in Figure 7.2 combines the Baddeley (1986) conceptualization of WM with the auditory versus visual modality dimension inherent to *multiple resources theory* (Section 7.3.3). However, both the WM model depicted in Figure 7.2 and Baddeley's WM model share the same assumption that is also central to CLT: that each of the WM subsystems is limited in capacity in the amount of cognitive processing that can take place at any given time.

In the updated model depicted in Figure 7.3, some of the aforementioned discrepancies in WM models are somewhat smoothed out. As indicated in Figure 7.3, in this conceptualization there is no demarcation between subsystems within WM. Instead, the assumption is that the various kinds of information that the learner encounters require integration

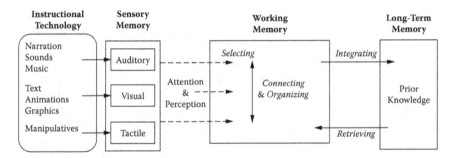

Figure 7.3 A cognitive theory of multimedia learning. (From Moreno, *Current Directions in Psychological Science*, 15: 63–67, 2006. With permission.)

and organization within WM, which is, as previously emphasized, subject to capacity and time limitations that will likely result in only a subset of the input information being further processed at any given time.

Figure 7.3 also provides some additional modifications that make it more compatible with the generic human information-processing model illustrated in Figure 7.1. First, it expands the scope of sensory memory to include the tactile sense, which can play an important role in many multimedia learning environments, adding another dimension of input information to which older learners need to allocate attention. Second, it more explicitly highlights the role of attention in the process of extracting information from sensory memory. In doing so, it invokes the many issues previously discussed in this chapter that could influence the learner's selection of information for further processing in WM. Also, the role of perception is acknowledged; as discussed earlier, mechanisms of perception can be affected by many factors that ultimately also can influence what information receives further processing in WM. Finally, in sharp contrast to Figure 7.2 but consistent with Figure 7.1, the model depicted in Figure 7.3 shows the dual pathway between WM and LTM. Thus, for the older adult, having more limited constraints associated with WM implies less meaningfully connected and organized information being transferred to LTM, and consequently a lesser ongoing accumulation of needed (prior) knowledge that can be retrieved from LTM as learning unfolds.

Referring to Figures 7.2 and 7.3, within the context of multimedia learning the potential for cognitive overload can arise from three kinds of cognitive demands: *essential processing, incidental processing,* and *representational holding*. Essential processing refers to cognitive processing related to selecting presentation material, such as words, images, and pieces of animations, as well as organizing and integrating this information for the purpose of making sense out of the instructional material. The importance of processing speed, which includes the speed with which the information selected can cue relevant schemas in LTM (i.e., perceptual speed) and the speed with which this information can be organized, make essential processing demands more challenging for older learners. Reduced WM capacity would also hinder essential processing.

Incidental processing is analogous to extraneous cognitive load in CLT. It refers to cognitive processing that is induced by virtue of the instructional task but does not contribute to making sense of the presented material. Representational holding refers to the cognitive effort entailed by holding selected information in WM over time so that it can ultimately be better understood or used for integrative purposes when other information becomes available. Instructional methods that can reduce incidental processing and representational holding, and redistribute essential processing so that cognitive processing resources can be used more optimally will reduce total processing required for learning and thus the potential for cognitive overload (Chapter 8).

7.5 Synthesis: Human information-processing system model with implications for older learners

Figure 7.4 depicts a model that summarizes a number of the insights into instructional design for older adults that were provided throughout this chapter from the perspective of human information-processing mechanisms. This model represents an integration of the following four models or frameworks that have been presented to this point:

- The generic human information-processing model presented in Figure 7.1, with an emphasis on the perceptual and cognitive stages
- Baddeley's (1986) model of WM that considers two capacity-limited WM subsystems, a phonological loop and a visuospatial sketchpad, and a central executive that coordinates these two memory subsystems
- Two of the dimensions associated with multiple resources theory, specifically: *processing codes* and its two levels, the verbal versus spatial code (which correspond, respectively, to the phonological loop and the visuospatial sketchpad subsystems); and *processing modalities* and its two levels, the visual versus auditory modality
- Concepts related to integrating diverse instructional materials in WM through mechanisms of attention and perception, the formation of verbal and pictorial models in WM, and the importance of retrieving information from LTM in support of WM integrative processes (Figures 7.2 and 7.3)

This integrated model highlights the often overlooked role of perception, especially in learning environments such as multimedia that may

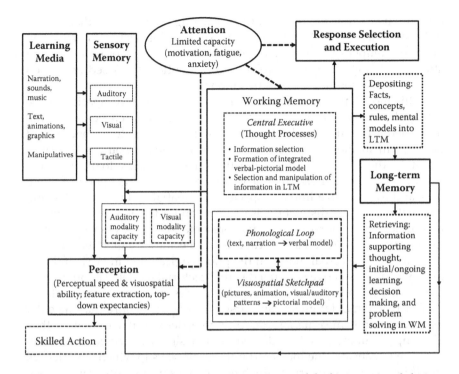

Figure 7.4 An integrative information processing model of instructional design with implications for older learners.

contain diverse instructional materials. Perceptual speed (the speed with which perceived information can be processed) is an ability known to decline with age, and instructional approaches that present material to the learner that demand rapid selection of information for closer examination risk having the older learner lose that material. Visuospatial ability is another concern. Some older learners, especially in light of the general slowing of information processing (Chapters 2 and 4), may need more time to accommodate pictorial elements. In fact, animations, as discussed in Chapter 10, may be problematic for many older learners, especially within the context of presentations that include other instructional media. Similarly, presentations of unfamiliar images that may require effortful bottom-up feature analysis can, as with other perceptual issues, demand resources of attention that could contribute to ongoing fatigue as well as reduce the ability to attend to other aspects of the instructional program.

Clearly, the older learner should not be put into a position of continuously having to decide what information to select. Instead, as indicated in Figure 7.4 by the arrow from LTM to perception, the older learner should be able to rely as much as possible on top-down expectancies based on knowledge in LTM to make such decisions. Reliance on knowledge accumulated about the topic and the learning environment that resides in LTM, which depends to some extent on how instruction is given, can offset or compensate to some degree for reduced perceptual speed and attention. Many other issues related to perception discussed in Section 7.3.1 are also accommodated by this model.

This model also points to the various ways that WM capacity can be manipulated, opening up the door for instructional approaches (including multimedia instruction) that could possibly compensate for age-related declines in memory. Consistent with the multiple resources theory of attention, the model provides two paths that can be taken to the presentation of instructional material: one that recognizes distinct pools of attention, and thus distinct limited capacities for spatial and verbal codes, and one that recognizes distinct pools of attention for the auditory and visual modalities. Thus, for example, the presentation of excessive narration information could overload the capacity of the auditory modality, leading to an ineffective verbal model in the phonological loop in WM, and ultimately to an ineffective integrated verbal-pictorial model. Although these model distinctions provide more ways to present instructional material that instructional designers can pursue in order to ensure that the older learner's overall information-processing capacity is not exceeded, they also impose an increased burden on these designers. Specifically, designers have to consider the increased number of ways in which a bottleneck in WM processing can arise, which now can occur not only from exceeding the visual or auditory modality capacity, but also by exceeding the verbal or spatial code capacity.

As emphasized in Chapter 4, progressing to higher stages of learning and skill acquisition requires the integration, organization, and conversion of information into rules, schemas, and mental models that provide, in addition to increased efficiency in skill, the ability to expand the envelope of knowledge into further learning and skill application. This can result from greater comprehension of the information and thus being more adept at reflecting on it, and from being better equipped to generalize the learning to new situations. Much of this ability depends on ensuring that WM capacity remains within the limits of the older learner during the ongoing instructional presentation, as discussed in detail in Section 7.3.2. Just as the role of perception, especially in terms of the processing resources of attention that it can consume (Figure 7.4), can be underestimated with regard to older learners with reduced cognitive capacity, the capacity required to organize and integrate information (Figures 7.2 and 7.3), whether it be from different processing codes (verbal and spatial) or different processing modalities (auditory and visual), may be a critical bottleneck for older learners obtaining instruction on complex cognitive tasks. There are many other implications of this model for instructional design for older learners that are implied throughout this chapter.

On a final note, and as indicated in the attention mechanism in Figure 7.4, we should not lose sight of the issue of overall reduced capacity of attention for many older adults, as well as factors related to motivation, fatigue, and anxiety (Chapter 6) that can indirectly affect the older learner's attention, and ultimately, all the critical mechanisms denoted in Figure 7.4 that are required for learning.

7.6   Recommendations

- When the continuity of learning depends on the rapid ability to locate critical stimuli such as buttons, tabs, or links, make these stimuli salient but not to the point of being distracting when not being utilized.
- Avoid having older learners invest large amounts of cognitive effort to obtain needed information as part of the learning process.
- Minimize the amount of divided attention required by the instructional program.
- If the instruction or training entails that the older learner needs to process information related to actions taken as well as the relatively immediate feedback of those actions, ensure that the amount of information in such instances is not capable of overloading the learner.
- If the older learner needs to take time to discern the properties of a stimulus in order to identify the entity properly (i.e., engage in bottom-up feature analysis), ensure that prior training emphasized these features or properties so that the learner has them available in LTM. This type of perceptual processing has a number of implications for the older learner.
- Be wary of any factor (e.g., poor illumination, masking noise) that is capable of degrading the stimulus characteristics. Critical features that are needed to attach meaning to the perception should be made available at a low investment of attention for the learner.
- To reduce the cognitive demands of perception, resort to task analysis applied to the instructional materials to identify where sets of symbols or cues can logically go together, perhaps pointing out to the learner that a set of information sources should be processed as a single entity. By unitizing material into chunks, the information processing load on the learner can be reduced.
- When training situations rely on the use of some form of display, adhere to the *proximity compatibility principle* (PCP) to ensure that the information contained within the display matches the demands of the task (see Section 7.3.1); older learners are more likely to be susceptible to mismatches in task–display proximity due to additional demands on attention.

- To ensure that any type of display is usable for older learners, minimize any visual or auditory *legibility* issue such as illumination, visual contrast, visual angle, noise, and masking.
- Avoid the need for the older learner to perform absolute judgment tasks during the course of training. If these tasks are necessary, restrict the number of levels associated with these types of judgments to about two to three (as opposed to the usual guideline of five to seven).
- When instruction requires older learners to move across various display sources to access information, locate information in such a way that minimizes the time or effort for learners to move their selective attention process from one display to another.
- During the course of learning, avoid imposing on older learners the need for retaining important information solely in their memory.
- Exercise caution regarding the amount of new information presented at any one time to the older learner and the rate at which that information is presented so that the learner has the opportunity to rehearse that material and move it into LTM.
- If instruction requires mixing verbal information (e.g., text, spoken instructions) with image-based (e.g., pictures, flowcharts) aim for a balance between these information sources rather than overloading on any one of these types of information.
- Take the opportunity to impart mental models to older learners to help them predict how a device or system may act under various circumstances and to be able to generalize this knowledge to new situations.
- Information that will need to be recalled for use later on during instruction should be rehearsed more strongly, and also rehearsed in combination with other information if the schemas of such associated information will be needed later on as well.
- Carefully plan the sequence with which information is presented, as this strategy can greatly minimize the demands imposed on WM (see Chapter 8).

Recommended reading

Bennet, K.B., Nagy, A.L., and Flach, J.M. (2006). Visual displays. In G. Salvendy (Ed.), *Handbook of Human Factors and Ergonomics*, 3rd ed. New York: J. Wiley, 1191–1221.

Casali, J.G. (2006). Sound and noise. In G. Salvendy (Ed.), *Handbook of Human Factors and Ergonomics*, 3rd ed. New York: J. Wiley, 612–642.

Cowan, N. (2001). The magical number 4 in short-term memory: A reconsideration of mental storage capacity. *Behavioral and Brain Sciences*, 24: 87–185.

Mayer, R.E., Griffith, E., Jurkowitz, I.T.N., and Rothman, D. (2008). Increased interestingness of extraneous details in a multimedia science presentation leads to decreased learning. *Journal of Experimental Psychology: Applied*, 14: 329–339.

McDowd, J.M. (1997). Inhibition in attention and aging. *Journal of Gerontology: Psychological Sciences*, 52: 265–273.

North, C. (2006). Information visualization. In G. Salvendy (Ed.), *Handbook of Human Factors and Ergonomics*, 3rd ed. New York: J. Wiley, 1222–1245.

Salthouse, T.A. and Maurer, J.J. (1996). Aging, job performance, and career development. In J.E. Birren and K.W. Schaie (Eds.), *Handbook of the Psychology of Aging*, 4th ed. New York: Academic Press, 353–364.

Sharit, J. (2003). Perspectives on computer aiding in cognitive work domains: Toward predictions of effectiveness and use. *Ergonomics*, 46: 126–140.

Sharit, J. (2012). Human error and human reliability analysis. In G. Salvendy (Ed.), *Handbook of Human Factors and Ergonomics*, 4th ed. New York: J. Wiley, 734–800.

Sweller, J. and Chandler, P. (1994). Why some material is difficult to learn. *Cognition and Instruction*, 12: 185–233.

chapter eight

Methods and approaches to instruction and training

8.1   Historical background

In Chapter 4, we discussed the general learning process as well as its close relationship to skill acquisition while also emphasizing the implications of these processes for older learners. Extending beyond these more general learning and skill acquisition processes is a long history of learning and instructional theories. These theories are rooted primarily in the current educational system, where psychologists have attempted to link theories of psychology to methods of instruction. Due to the number of diverse perspectives that have dominated the science of psychology from the early 1900s to the present time, tracking the influences of these views on instructional design is clearly beyond the scope of this book. It would also lead us astray from the emphasis on training older adults, as many of these psychological theories of learning were driven by the need to establish curricula in formal education.

Moreover, a number of the constructs, concepts, and methods underlying these approaches to instructional design are addressed, in one form or another, throughout this book. For example, the roles of simulation, interactivity, and virtual reality in learning are discussed in Chapters 5 and 10, and the use of instructional systems design for the analysis of instructional content and the organization of training programs is discussed in Chapter 9. In the present chapter, methods related to task analysis, the sequencing of instruction materials, and the use of holistic tasks during instruction are emphasized. However, before we briefly present some of these and other ideas related to instructional design, it is useful to provide a very brief historical account of the kinds of thinking in this area that have led to current perspectives in instructional design (ID).

R. D. Tennyson (2010) referred to John Dewey, who was instrumental in establishing the idea of linking learning theory to educational practice, and Edward Thorndike, who researched principles of learning that could be applied to the teaching process, as the two primary theorists who were instrumental in laying the foundations of ID. B. F. Skinner, the well-known proponent of behaviorist theories, applied these theories to a framework of instruction that includes small incremental steps, sequencing from simple to more complex material, learner participation, reinforcement of correct responses, and individual pacing.

Despite the leaning away from behaviorist learning models and theories and toward cognitive learning theories in the late 1960s and early 1970s, the behaviorist influence is still very much evident in many current ID models or approaches. For example, the pivotal roles of sequencing, not overloading the learner, promoting active participation, and the provision of feedback are all elements that comprise essential components of today's ID models, and especially those that are tailored toward older learners. In addition, these learning theories and models influenced many other significant contributors to the field of learning and ID, including E. D. Gagné (1978) (Chapter 5).

The transition to cognitive-based frameworks added some new ideas that expanded the scope of ID. For example, Bruner (1966) believed that as the learner develops expertise on a topic, ideas should correspondingly be reintroduced in increasingly complex ways. This notion is somewhat analogous to Ausbel's (1969) theory of *progressive differentiation*, which proposes that general ideas should be followed by a sequence of more concrete information. These perspectives are very consistent with ideas related to sequencing instruction, which is elaborated on in this chapter. The ideas generated from these perspectives are very relevant to ID for older learners, as they serve to minimize an emphasis on working memory early on in instruction (Chapters 3 and 7), and instead focus on establishing relevant schemas on which the older learner could subsequently build. Bruner has also been very closely associated with the concept of *discovery learning* (Section 8.8.2). Cognitive learning theories have also focused on the development of ID approaches for improving concept formation, which remains a very active area of research.

The cognitive perspective was evident as well in the tendency to apply *task analysis* (a hierarchical analysis of the steps needed to perform a task) and *content analysis* (defining the facts, concepts, principles, and procedures corresponding to the subject matter and how these attributes are organized relative to one another) as a basis for identifying different levels of learning in ID. However, as the ID area began to embrace new learning methods such as computer-based simulations, it became evident to some ID researchers that these task and content analysis procedures were more suited to passive ID methods, but did not lend themselves well to highly interactive instructional situations that involve interactive media.

To account for how information is integrated into meaningful wholes in order to understand complex and dynamic phenomena, cognitive structures referred to as schemas began to receive much greater recognition. At the same time, the roles of attention, perception, and memory were also accorded much greater importance in skill acquisition (Chapters 4 and 7).

Earlier, we alluded to perhaps the most fundamental and critical aspect of instruction, namely the issue regarding the sequence with which information should be presented to a learner. This remains an essential topic within the cognitive perspective to learning theory and ID, and has enormous implications for ID directed at older adults. The assumption is that better organization of information in memory would lead to more reliable access of that information, thus enabling higher cognitive activities such as problem solving and even creativity. The organizational structure of this information within the learner's knowledge base for given problem contexts, as well as the concepts that the learner may invoke to solve problems (i.e., the individual's internal organization of information), are often inferred by analyzing a problem's complexity (i.e., the external structure of the learning material) and how individuals try to solve those problems. It is believed that these analyses can lead to better sequencing in the presentation of new information. In principle, however, a *problem context* and its corresponding contextual knowledge include not only the content and task analysis information referred to earlier, but also cultural aspects (e.g., the appropriateness of the information) associated with a given situation. This concept is at the root of what is commonly referred to in the field of learning as *situated cognition* (Section 8.8.3).

Many other concepts related to cognition and cognitive science subsequently became incorporated into instructional theory and ultimately into frameworks for ID. These became reflected in approaches to ID that were directed at achieving specific, well-defined performance outcomes that were based on assessing a learner's progress and thus establishing the learner's needs, from which instructional strategies, sequences, and media could then be determined. The idea of a single, most effective approach to all instructional situations was largely abandoned. Instead, numerous ideas and theories emerged. There was also increased emphasis given to the learner's production of knowledge, including strategies for enabling learners to organize information, reduce anxiety, and develop self-monitoring skills. Closely related to this concept was the idea of *metacognition*, where learners are aware of their own cognitive abilities and the strategies they use to acquire and utilize knowledge. An overview of many of these and other historical trends in learning and their influence on instructional design can be found in Tennyson (2010).

In this chapter, we present a number of relatively current ideas associated with training and ID methods that we feel are particularly relevant to older learners. Specifically, we begin with an emphasis on methods that address the issue of how to sequence instructional materials to the learner. This is an area that has received a good deal of attention in the cognitive science domain, but which also has broad links to many other aspects of instruction that we touch upon throughout this book that concern older learners. We then briefly present a number of other methods, with coverage of the topics of e-learning and multimedia largely set aside to Chapter 10.

8.2   The importance of sequencing in instructional design

Instruction generally involves two broad decisions: what to teach, including tasks, skills, and higher-order thinking skills, which are often referred to as *scope decisions*; and how to group and order the content of the learning material, which are often referred to as *sequence decisions*. Assuming that we have a reasonably good idea of the domain content that we want to teach, the focus is then on how we should group and sequence this material appropriately. Generally, sequencing is considered to have a potentially important impact on learning when there are strong relationships among the elements of the content and when the amount of content is reasonably large (e.g., requires at least an hour of instruction). Moreover, the types of relationships inherent in the content can have a strong influence on the types of sequencing strategies that will be most effective.

When the training will encompass several topics, there are two basic patterns of sequencing to which the various sequencing strategies can be applied: *topical sequencing* and *spiral sequencing*. In topical sequencing, a topic or task is taught to some target level of depth of understanding or competence before training on the next task is initiated. In spiral sequencing, the learner achieves mastery of several interrelated tasks by passing through all the tasks at a basic level, and then continues to pass through these topics again, in the same sequence, each time at a greater level of depth, until the target level of learning is achieved for all the tasks.

The advantage of topical sequencing is that the learner can focus on one topic without the possible disruption by learning associated with other tasks. The disadvantages are that previous material can become easily forgotten when the emphasis shifts to a new task, and an appreciation for the whole subject domain may have to be deferred until the training is completed. Carefully choosing points for reviewing and synthesizing prior learned material could, however, mitigate some of these negative aspects. Spiral sequencing, in contrast, offers the possibility for learning interrelationships among tasks more easily as it provides close temporal contiguity with regard to aspects of the various tasks that are similar, and also a periodic review by virtue of cycling back to learn an earlier topic in greater depth. However, despite its built-in synthesis and review features, spiral sequencing can disrupt the process of schema development by learners who are trying to develop a model of the task.

In reality, many types of learning are probably best accomplished somewhere in between the two extremes of pure topical or spiral sequencing. Determining where in the continuum between topical and spiral sequencing training should be directed is likely to be influenced by how deeply the instructor feels the task needs to be pursued before transitioning to another task.

Before discussing the various sequencing strategies and their implications for training older adults, it is important to emphasize that the distinctions between the two basic patterns of sequencing, topical and spiral, can have important implications for the role that sequencing can play in older adult learning. Assuming that older adults are more susceptible to disruption in the development of organized knowledge about a topic (i.e., to schema development; Chapters 4, 5, and 7), it would seem that spiral sequencing would present greater risks to learning by older adults. If that is the case, when tasks are interrelated it is probably safer to identify points where the learner appears to have attained a reasonable mastery of the particular task or topic. Instruction should then consider reviewing a prior learned task first in isolation to strengthen the traces associated with that task. When sufficient mastery has been achieved, links to related aspects of the new task(s) can be established in order to strengthen associations between the tasks.

The idea that learning a task or topic should be influenced by the order in which the topic's elements are presented is intuitive. Consider the frustration of trying to piece together a story or information that is presented out of sequence. The mental effort required to establish some sort of model or schema from the information can be overwhelming. In some informal learning situations, for example, during the passive viewing of a suspense film, information may be intentionally provided out of sequence in order to induce within the viewer or learner uncertainty and confusion, perhaps to heighten the suspense. However, this burden of uncertainty is not desirable in practical situations where, unlike the passive role often assumed in entertainment situations, the learning is taken more seriously and is needed to develop essential or desired knowledge and skills.

Imagine the older learner in a situation where information is not appropriately sequenced. The information-processing effort required to make sense of the material could be very excessive, and coupled with possibly existing doubts concerning one's cognitive capability and task domain-specific efficacy, could result in frustration and learning failure. The ordering of information presented to the older learner is thus critical. In the ensuing sections, strategies for sequencing instructional material to be learned are presented. The strategies should not change for the older learner; however, aspects of these strategies may need to be given different degrees of emphasis in consideration of the challenges many older adults may face when confronting new tasks or topics. These strategies and some of their implications for older learners are summarized in Table 8.1.

Table 8.1 Sequencing Strategies in Training and Instruction

	Definition	Implications for Older Adults
Basic Sequencing Patterns		
Topical sequencing	Training is directed at achieving a target level of competence on a topic or task prior to training on the next topics.	This strategy benefits older adult attentional processes by enabling them to focus on one topic at a time. The strategy is especially suitable if the tasks can be largely decoupled from one another. If performance requires synchronization of the topics, carefully choose points along the way for reviewing prior topics and synchronizing them.
Spiral sequencing	Training is directed at achieving a target level of competence on several interrelated tasks or topics prior to presenting additional passes through this set of tasks to achieve higher levels of competence.	Use this approach with older adults only if the tasks or topics comprising the target instructional goal are highly interrelated. If this approach is used with older learners, ensure that the first pass through the tasks begins at a very fundamental level of skill and understanding.
Specific Sequencing Strategies		
Hierarchical sequencing	A hierarchical task analysis (HTA) is used to identify a systematic ordering, from low to high, of prerequisite skills needed to achieve the training goal. The initial focus in training is on skills related to being able to discriminate task elements from one another. This enables training on concepts (concrete or defined), which in turn serve as the basis for training on rules (procedural or heuristic). HTA also supports the linking of these skills to their corresponding task components.	This strategy is very suited to training older adults as it emphasizes the identification of prerequisite skills, which in many learning situations, especially those involving new technologies, many older adults may lack. Lower-level prerequisite skills should be emphasized and not assumed to be available to the older learner. This strategy, however, does require instructional designers and trainers to be skilled in performing an HTA. It also does not provide any guidance in how to link skills at one level of the skills hierarchy to skills at higher levels, which may demand integrating other sequencing strategies with this one.

(continued)

Table 8.1 Sequencing Strategies in Training and Instruction (continued)

	Definition	Implications for Older Adults
Procedural sequencing	For tasks whose components are comprised of procedural steps that have well-defined orders (i.e., a particular step needs to be performed prior to a subsequent step) an HTA should be performed but one that focuses on the requisite ordering of procedural steps rather than skills.	For older learners, consider the level of detail that is being taught concerning each procedural step to ensure that too much detail is not imparted on any given procedural step early in training. If the number of steps is not excessive, consider incorporating a spiral sequencing whereby passes are made through a sequence of procedural steps at progressively greater levels of detail.
Elaboration sequencing	The sequencing strategy will depend on different kinds of expertise that the instruction is intending to develop. A distinction is made between task expertise (expertise on aspects of performance specific to a task) and domain expertise (broader, more generalizable expertise that can be applied to a task domain).	Elaboration sequencing directed at developing task expertise focuses on cognitively complex tasks that are often performed differently under different conditions. This can be especially challenging for older adults as the expertise includes recognition of the different kinds of conditions (e.g., different steps or different principles) that serve as the basis for action or decision making.
The simplifying conditions method (SCM) of generating elaboration sequences for task expertise	For both procedural and heuristic tasks, the SCM requires identifying the simplest versions of the task that are still whole and, one hopes, real-world representative versions of the task. These holistic task versions are then systematically increased in complexity following mastery on prior versions. With procedural tasks, the increases in complexity require that additional conditions be considered. With heuristic tasks, the complexity of the principles and causal relationships underlying these tasks are systematically increased.	Subjecting the older learner to very simple, yet holistic task versions can facilitate the establishment of rudimentary cognitive schemas. These schemas, in turn, can help these learners absorb more complex but related material as instruction progresses. This could increase the chances of successful training on such complex tasks, partly by reducing the overall cognitive load of learning by not placing demands on forming complex schemas from previously accumulated and possibly more fragmented task-specific knowledge. For heuristic tasks, consider using problem-based simulations as the task versions become increasingly more complex in order to aid the learner in the discovery of principles.

(continued)

Table 8.1 Sequencing Strategies in Training and Instruction (continued)

	Definition	Implications for Older Adults
Conceptual and theoretical elaboration sequences for domain expertise	A conceptual elaboration sequence derives from a conceptual analysis that first considers the broadest, most inclusive concepts associated with the domain and progresses to narrower, less inclusive elements until a target level of refinement is reached. The result is an often hierarchical taxonomic structure that is capable of providing answers to a variety of "what" types of questions. A theoretical elaboration sequence is used for domain expertise more focused on the "how" and "why" concerning objects or processes. The sequence begins with the broadest, most inclusive general principles and then progresses to more detailed and complex knowledge.	The development of conceptual and theoretical elaboration sequences for older learners requires caution as concepts and principles can be inherently abstract and thus become easily confused, resulting in ineffective learning. In generating elaboration sequences that are top-down, whether in terms of concepts or principles, as the elaboration sequences progress downward, links to higher levels should be periodically reinforced to lessen the possibility for confusion of concepts and principles. Caution must also be used with regard to the amount of information that is presented in a session, and short rest breaks should be considered with such elaboration sequences to promote reflection of the concepts and principles at any given level of detail and their top-down associations.

8.3   Strategies for sequencing the order of instruction

8.3.1   Hierarchical sequencing

Many tasks have certain types of knowledge or skills that need to be learned before others can be learned. If these prerequisite skills can be identified, the hierarchical sequencing strategy dictates that these skills be taught first. For example, it would not make sense teaching someone about scrolling on a computer before introducing the concepts of a mouse and cursor.

The sequential ordering of the prerequisite skills and knowledge can be accomplished using a *hierarchical task analysis* (HTA). This task analysis method is used extensively in human factors applications to identify and subsequently assess the physical and cognitive demands that people face when carrying out various tasks. In HTA, the task the user needs to perform is decomposed hierarchically into goals, the plans for meeting these goals, and the operational steps for carrying out these plans. In Chapter 15 of Fisk et al. (2009), it is shown how HTA, combined with other tools, can be used to predict problems older adults might encounter when interacting with a product or system.

When used for establishing the proper sequence of training or instruction, HTA can benefit from a general hierarchical ordering of skills from the bottom upwards. These skills can then be mapped into their specific elements for a particular task. At the bottom of such a hierarchy are very fundamental skills such as the ability to discriminate stimuli from one another along one or more physical dimensions, which requires some degree of generalization. For example, on a computer display the artifact referred to as a *dropdown box* may need to be differentiated from the artifact referred to as a *window*, and differences between different dropdown boxes should be able to be differentiated as well.

This ability is essential for learning *concrete* and *defined concepts*, the next two higher levels, respectively, of intellectual skills, although the distinction between these two concepts is often very subtle. To obtain the skill for a concrete concept the learner needs to identify a stimulus as a member of a class having some observable property. Thus, rather than just being able to state that a dropdown box is different from a message window, the person can now identify any type of dropdown box as belonging to a class of artifacts; that is, the learner develops the skill by generalizing from instances. Defined concepts enable the learner's performance of the skill to be guided by a definition of a property that all members of such a class have in common.

Concepts, whether concrete or defined, are used in turn to train the learner to formulate rules, which define classes of relationships among classes of objects and events. For example, a rule built on various concepts may be, "If an arrow is produced when hovering over an item in a sidebar directory on a homepage of a website, then moving laterally produces a dropdown menu associated with that item." Another rule may be, "If the claim is based on damage that occurred on commercial property and the damage is confined to flooding, and the claim does not exceed $10,000, classify the claim event in the database as minor." Both of these rules would be considered *procedural rules* as they specify the conditions for identifying some state or performing some action. Training may also involve instruction on the use of more general rules, sometimes referred to as *heuristic rules*. These rules are usually expressed in the form of principles and guidelines, for example, "Never process claims that lack an identifying code," or, "If you are uncertain how to return to a previously viewed webpage keep hitting the back button." Defined concepts are, in a sense, rules themselves; however, they would be considered *classifying rules*, which are often needed to instantiate a procedural rule.

Attaining an increased skill level on a task will often involve combining classifying, procedural, and heuristic rules. This process establishes higher-order rules that are more complex than their constituent rules, and which form the basis for the kinds of problem-solving activities required for achieving performance goals.

The hierarchical sequencing approach assumes that the training sequence does not teach a skill before the prerequisite skills have been addressed. It also assumes an understanding of the learner's starting knowledge. Because different learners bring to the instructional setting different knowledge and skills, ascertaining an appropriate starting point for instruction is critical. With older adults in particular, no assumptions should be made concerning even basic skills, especially technological skills, and it is well worth the effort to establish a baseline level of knowledge that the older learner has, even about skills that may be ancillary to the tasks being taught.

An HTA approach to training, however, does not impose constraints on how instruction through the hierarchy should be delivered. Thus, with a spiral (i.e., depth-first) approach to hierarchical sequencing, all the skills at the lower level would be taught prior to proceeding with instruction at the next higher level. In contrast, with a topical (i.e., breadth-first) approach, instruction would focus on one topic, starting with the lower-level skills on that topic, and then proceeding to higher levels within the HTA on that topic prior to addressing a different topic in the hierarchy. Other sequencing strategies within the HTA are also possible.

It must be noted that not all instruction involves the sequencing of skills in which one skill is linked to another. When the leap between the skills being taught is relatively broad, the hierarchical sequencing approach may not be appropriate as it does not offer guidance on how to sequence such skills. With older adults, who may require a greater degree of decomposition of concepts and rules during sequencing, a hierarchical sequencing strategy may impede the formation of schemas or more complex organizations of knowledge by forcing an emphasis on fragments of knowledge and skills. However, despite the realization that the sequencing requirements of learning are not likely to be satisfied solely by a hierarchical approach, this approach does help ensure that training averts the kinds of performance failures that would otherwise be almost guaranteed, especially for older learners, when such sequencing is violated. The true power of hierarchical sequencing lies in combining it with other sequencing strategies, which are presented below.

8.3.2 *Procedural sequences*

Procedural sequences are based on the idea of a *procedural prerequisite*, that is, a step during the execution of a task that must be performed before another step can be performed. A procedural prerequisite is distinct from a *learning prerequisite*, which implies the need to learn some skill, for example, a concept, before another skill can be learned. The instructional strategy is then to perform an analysis of the steps that comprise a procedure and to identify the corresponding procedural prerequisites in the order of their performance. In a *procedural task analysis*, in essence one performs an HTA, but one that is directed toward the decomposition of procedural steps rather than the decomposition of the broader idea of skills.

Two points should be noted that have potentially important implications for training older adults. First, as with HTA, any step (or skill) that is decomposed is intended to continue to provide a full description of that step, but with more detail. The assumption is that more detail is conducive to more effective learning because critical elements of information that may be prerequisite for other elements are less likely to be absent. However, there is still the issue of sheer quantity of information to consider. An increase in quantity of information in a procedural sequence can also arise from procedures that are replete with decision steps and branching points (as is typical of many task analyses developed for skilled industrial work activities), especially when it is not obvious how instruction on such branching should be accomplished.

Assuming that the presentation of such large amounts of information can jeopardize one's learning, including that associated with the appropriate sequencing of tasks, an issue that can easily arise with older learners is whether presenting a relatively large amount of material concerning individual task steps might result in decreased motivation or be perceived as overwhelming and thereby lower the learner's level of focused attention. If so, then instruction that decomposes the material into more manageable units while preserving the emphasis on sequencing of steps should be considered. To accommodate the greater amount of detail related to task steps that needs to be absorbed, instruction can be divided into smaller sessions separated by more rest breaks (Chapter 4). Spreading out the learning across more sessions, however, may result in some learning being lost, which may necessitate adding a degree of refresher learning to bridge across the points within the procedural sequence where instruction was temporarily halted. These are typical of the kinds of tradeoffs that instruction directed at older adults needs to consider.

A second point related to procedural task analysis is that any given procedural step, and especially those at the more entry (lower) levels of the analysis, may entail the need for training on concepts that are critical for performing the procedural step. Consequently, a procedural task analysis at particular procedural steps may have associated with it an HTA that is directed at explication of needed concepts for that step. For example, in the process of providing instruction on how to search a company's database for information, instruction may be required concerning the distinction between different types of records comprising a portion of the database. An HTA could then be used at this point in the procedural sequence to elaborate on the concept of records. In principle, the procedural sequence strategy can be combined with other types of sequencing strategies as well.

8.3.3   Elaboration theory and elaboration sequences

The *elaboration theory of instruction* is based on the idea that the sequencing strategy may depend on different kinds of expertise that the instruction is intending to develop. Specifically, different kinds of expertise may depend on different relationships that exist within the content of the material to be learned and thus may require different sequencing strategies (Reigeluth, 2007). This theory distinguishes between two broad types of expertise: *task expertise* and *domain expertise*. Task expertise relates to the learner becoming proficient in a specific task such as filing insurance claims, using Excel to create a spreadsheet, or finding and correcting problems with products. In contrast, expertise in subject matter that is not linked to a specific task would constitute domain expertise. Such expertise would include understanding the basic menu structure associated with all Microsoft business applications or fundamental concepts associated with insurance claims.

When the elaboration theory of instruction is directed at developing task expertise, the focus is generally on cognitively complex tasks that typically need to be performed differently under different conditions. The simplifying conditions method (discussed below) of elaboration theory emphasizes simple-to-complex sequencing of holistic versions of the task. Thus, rather than focusing on a parts-to-whole sequencing approach, this theory of instruction focuses on the presentation of an actual task at different levels of complexity, beginning with the simplest version of the task and progressing to more complex versions when proficiency is demonstrated on prior task versions. Cognitively complex tasks can include tasks that are primarily *procedural* (i.e., steps are used to decide what to do when) or *heuristic* (i.e., interrelated sets of principles or guidelines are used to decide what to do when).

The key point is that the learner is presented with holistic tasks in a progressive simple-to-complex sequence. As shown further on, this idea is one of the most critical attributes of more recent instructional design models such as the four-component instructional design model (Section 8.5), which, building on this fundamental strategy, offers other features for ensuring effective learning. It does, however, place the burden on instructional program designers to develop alternate versions of holistic tasks, which can be a challenging creative enterprise. Also, and especially relevant to the older learner, it entails designing an "entry point" whereby the learner has a reasonable opportunity to deal successfully with the task with some help.

With instruction directed at establishing domain expertise, where the nature of the expertise can range from general to detailed, the primary approach to sequencing derived from elaboration theory is to begin with the broadest, most general ideas and progress to more complex, precise ideas. The theory considers two major types of domain expertise: conceptual expertise and theoretical expertise. The former is directed at knowledge intended to help understand the "what," whereas the latter is directed at knowledge intended to help understand the "why." Although elaboration theory offers approaches for developing holistic, general-to-detailed sequences of instruction for gaining task or domain expertise, learning may involve both types of expertise, implying that both types of elaboration sequences could be used simultaneously. However, this strategy, which is referred to as *multiple-strand sequencing*, should be used with caution with older adults as it could impose too large a cognitive load.

In principle, there may be a variety of different types of elaboration sequences. Three types of elaboration sequences of instruction described below correspond to the three types of expertise that have been noted: the *simplifying conditions method* for task expertise (procedural and heuristic), the *conceptual elaboration sequence* for conceptual domain expertise, and the *theoretical elaboration sequence* for theoretical domain expertise.

8.3.3.1   Task expertise: The simplifying conditions method

With the simplifying conditions method (SCM) of generating elaboration sequences of instructional material, the focus is on generating simple-to-complex sequences of holistic tasks. This is in contrast to the hierarchical sequencing approach, which sequences instructional material based on fragmenting a task into its logical ordering of prerequisite component task skills and teaching those component skills first prior to instruction on a complete version of the task.

For both procedural and heuristic tasks, the SCM requires identifying the simplest version of the task that is still a whole and, it is hoped, representative, version of the task. The assumption is that most complex tasks have, under certain restricted conditions, simple or primitive versions of their more complex counterpart versions. The subsequent task versions that are presented to the learner should continue to be whole versions of the task but with systematic increases in complexity as prior versions are mastered. With procedural tasks, the increases in complexity will likely require that additional conditions be considered, which will result in more decisions, branches, and paths to negotiate.

With heuristic tasks, the SCM sequence is based on gradually increasing the complexity of the principles and causal relationships that underlie the skills associated with these tasks. If explicit elaboration of such knowledge at increasingly complex levels is difficult, the use of problem-based simulations can be used to capture increasingly more complex scenarios from which principles can be discovered (Chapter 10). Generally, tasks require a combination of both procedural and heuristic knowledge, in which case the SCM sequence can be used simultaneously for both.

8.3.3.2 Domain expertise: The conceptual and theoretical elaboration sequences

Consider a sales job that requires a worker to learn about the various types of marine vessels and associated equipment that a company sells. The conceptual elaboration sequence should begin with the broadest, most inclusive concepts (e.g., three broad classes that all the vessels fall into) and then progress to narrower, less inclusive elements until a target level of refinement is reached. The various concepts can be derived from some form of conceptual analysis capable of producing a corresponding conceptual or taxonomic structure. These structures are often hierarchical in nature, implying that the conceptual elaboration sequence is consistent with hierarchical sequencing. The conceptual structure, however, should be sufficiently informative to provide answers to a variety of "what" types of questions, for example, "What type of boat should we buy for seven-day trips to the Caribbean and what types of accessories are most critical?"

In contrast, the theoretical elaboration sequence is used for domain expertise that is more focused on the *how* and *why* concerning objects or processes. For example, the interest may be in training workers on how to interface with a largely automated system that controls a pharmaceutical mixing process. The worker would need to know how the process works, including which functions are performed by automation and which are controlled by the human, and why certain actions are needed. Typically, this type of domain expertise requires knowledge concerning interrelated sets of principles at different degrees of detail.

As with the conceptual elaboration sequence, the theoretical elaboration sequence begins with the broadest, most inclusive general principles. In the case of the process control example, these principles might address the purpose of the process and the fundamental inputs and outputs of the process. The sequence then progresses to more detailed and complex knowledge. Continuing with the pharmaceutical process control example, such knowledge may include basic thermodynamic principles underlying the detection of excessive heat and strategies for controlling various perturbations.

8.3.4 Elaboration sequences: Implications for older adults

Elaboration sequences have some attributes that make them potentially very useful as a basis for providing training to older adults, regardless of whether the training is done face-to-face or is computer-based. Perhaps the most critical attribute of such sequences is that, by enabling the learner to encounter a holistic task from the start, a cognitive schema can be developed that could facilitate the absorption of more complex but related material. This could ultimately serve to reduce the overall cognitive load of learning on the older person and increase the chances of successful training.

Gradually increasing the complexity of holistic tasks should also increase the likelihood of learning complex tasks, as this strategy would enable more complex concepts to be more easily absorbed. Holistic tasks are also more suited to on-the-job training, problem-solving, and the use of computer-based simulations, providing a greater range of flexibility for training older adults. The ability so early on in training to perform a holistic task could also have a powerful motivational effect on older learners, giving them the confidence that they can perform the complex tasks, and perhaps increasing their focused attention to task versions that are subsequently presented to them. In addition, encountering more difficult problems earlier in the block sequence could lead to frustration and negative emotional states, which could be exacerbated in older adults who may already be dealing with lower self-efficacy (Chapter 6).

Designing the kinds of holistic tasks that elaboration sequences demand can, however, be a very resource-intensive process for instructors, and there is no guarantee that the sequencing, especially if not formulated well, will not produce confusion, especially in older adults. Thus, if the instruction is targeting a number of small unrelated topics, it is probably not worth the effort to construct sequences of holistic tasks.

If the training exercise is very large in scope, it may be necessary to segment the training into episodes or *blocks*. Otherwise, the ability to recall earlier learning experiences within a large training episode as the learners attempt to absorb what they have recently learned can become compromised. The other extreme, fragmenting a relatively large training program into episodes that are too small, is also potentially problematic as this strategy may incur an overhead associated with having to bridge the gaps between episodes through repetitive refreshers of material. Such a strategy may also induce a loss of motivation in the learner.

Because fatigue is an important consideration when training older adults (Chapter 6), the trainer should probably take a conservative posture with regard to how much material should be presented in an episode or block of training. The costs of fragmentation of episodes are likely to be outweighed by the costs attributed to fatigue and the concomitant loss of sustained attention that might occur across the course of a longer instructional episode. The creation of a larger number of learning blocks will impose the need to "link" these episodes to maintain continuity in learning; the training program should thus have a feature that always reinforces or summarizes the essence of the just completed block(s) of tasks and provides a logical lead-in to the upcoming learning episode.

This will also help the older learner categorize the episodes and thereby aid in associating rules or needed knowledge with different scenarios. For example, one block of tasks may deal with tasks that address variations in how to locate different products in a company's product database, whereas another block may focus on how to update the database in the event that products have been discontinued, added, or will be coming out in the near future.

8.4   Sequencing worked examples and problem-solving exercises

There are many training and instructional settings that rely on the presentation to learners of worked examples. In fact, many academic learning situations are heavily dependent on this approach to skill acquisition, whether they are in the form of textbook, computer-based, or instructor-based demonstrations of worked examples. The presentation of worked examples presumes that at some point later on the learner will be transitioning to problem-solving activities that serve as reinforcement and extension of the principles underlying the solutions in the worked examples.

Although many of the types of learning situations for which this approach to skill acquisition is used represent classic academic subject topics such as mathematics and physics, worked examples and problem-solving exercises also apply to a wider array of learning situations. In these situations, worked examples can serve as a sample template for the types of problems that trainees may encounter and will need to solve. In the ensuing section, initially we examine the importance of example-based learning for skill acquisition and properties of worked-examples instruction that could be conducive or detrimental to learning. The focus then shifts to the issue of how to order these worked examples and the subsequent problem-solving instructional elements, and the possible implications of these ordering decisions for older learners.

8.4.1   Example-based learning

Skill acquisition theories typically distinguish between its different phases, for example, early, intermediate, and late. However, as was noted in Chapter 4 the boundaries between these stages are far from clear, especially when the cognitive skills being acquired are complex. Thus, if there are many different aspects of knowledge involved in the learning of a particular domain, the learner may be simultaneously in different states of skill acquisition with regard to these different aspects.

In the early phase, the learner typically is not trying to apply any of the knowledge that is being presented. Rather, the learner is trying to assimilate the various elements of the learning materials in order to gain a basic understanding of the knowledge domain. Well-designed worked examples presented during this phase of learning thus have the potential to emphasize important underlying principles. During the intermediate phase, the learner will try to abstract the principles from these worked examples in order to apply this knowledge to solving concrete problems. In the process, this phase will serve to highlight flaws in one's knowledge base, usually in the form of facts, rules, relations, or other aspects of cognition related to understanding of the knowledge, and thus provide the learner with the opportunity for correcting these flaws. In the later phase of learning, practice serves to facilitate speed and accuracy in problem solving.

The ability for the learner to correct flaws in knowledge resulting from worked examples provided during the early stage of learning is, however, not a process that is ensured simply by examining the solutions to these worked examples. What researchers have found is that learners need to actively self-explain or reason about the rationales for their solutions. Another important strategy is to pair worked examples with practice problems. This encourages reflection on the part of the learner, which facilitates self-correction of flawed knowledge and thus skill acquisition.

Generally, what makes studying worked-out examples so effective is that the cognitive load that these examples impose on the learner is mostly *effective load* (Chapter 5). This implies that the learner's information-processing activities are directed at reflecting on solution procedures in order to gain an understanding of how to apply domain principles to various types of situations. If early in instruction the emphasis instead were on solving a problem, the learner's cognitive load could be largely *extraneous*. In this case, the learner may be compelled to allocate significant resources of working memory and attention (Chapter 7) toward formulating strategies for solving a problem, leaving little spare cognitive processing capacity that could be applied toward reflection.

However, once learners gain some degree of expertise in the knowledge domain, instruction through problem solving has been found to be more effective than instruction in the form of studying worked examples. This finding suggests an instructional sequence effect across phases of skill acquisition. In the later stages of skill acquisition, reflection of worked-out solutions through self-explanations may actually be an impediment to learning. Instead, the focus should be on practice involving solving problems to promote speed and accuracy.

The role of self-explanations in the use of worked-out examples has important implications for training older adults. As noted, a key to the effectiveness of worked-out examples is their ability to promote active reflection through self-explanations (refer also to Chapter 10 on designing worked examples from the perspective of e-learning). In particular, self-explanations should foster the identification of the underlying knowledge domain principles, and for certain types of problems they should demonstrate how certain operations achieve subgoals that can describe the goal structure of different problem types. These types of reflection activities are especially effective when the learner has a low level of prior domain knowledge, which is often the case with older learners, for example, when they are expected to learn new technologies. Experiments that have explicitly trained people to generate self-explanations have been found to produce better learning outcomes (e.g., Renkl et al., 1998). These studies used a variety of different methods of explicit training in the application of self-explanation strategies such as providing information on the importance of self-explanations, coached practice with self-explanations, and encouraging self-explanations.

Another strategy that has been found to be effective once some knowledge in the early stage of skill acquisition has been acquired is to present worked examples, but with some blanks that the learner has to fill in so that the learner has to anticipate the next solution step on his own. This strategy, coupled with the provision of feedback on the correct step following the anticipation, could improve the utility of using worked examples during instruction.

One concern with self-explanations, and in particular relying solely on learners being able to provide self-explanations, is that the learner may not be able to provide such explanations, even with coaching, or may provide incorrect self-explanations. For the older adult, providing an incorrect self-explanation may be more risky as older adults may be less capable of recovering from establishing false principles or relations. This raises the issue of offering *instructional explanations* as opposed to promoting self-explanations. However, designers would need to be aware of the possibility that the content implicit to instructional explanations may overwhelm the cognitive capacities of many older learners, especially at the early stages of skill acquisition.

By carefully considering the nature of the instructional explanations and the order in which they are used within the instructional program, the potential exists for combining self-explanations with instructional ones to create effective learning outcomes. In fact, when used effectively, instructional explanations could support spontaneous self-explanation activities. Some of the principles developed by Renkl (2002) for the design of such instructional explanations follow.

- To ensure that instructional explanations are appropriately timed, present them when the learner requests them, or in reaction to a learner's error when filling in a blank in a worked example when the learner is attempting to anticipate the next solution step. This will increase the likelihood that the instructional explanation will be used by the learner to gain an understanding of the domain knowledge.

- Make the instructional explanations concise so that they have the best opportunity for being processed. This is especially beneficial to older learners who are more susceptible to information overload.
- Ensure that the content of the instructional explanations focuses on underlying principles of the topic.
- Link the instructional explanation as much as possible to the worked example. For older adults, this provides a kind of consistent mapping, whereby external explanations (e.g., from face-to-face instruction or from a computer-based assistant) are always integrated into the example, reinforcing the knowledge associated with the content of the example.

Overall, these principles address the transition in example-based learning from the use of worked examples to problem-solving exercises as the learner transitions from the early to the later stages of skill acquisition. This transition is encapsulated in the following instructional order model proposed by Renkl and Atkinson (2007), and depicted graphically in Figure 8.1.

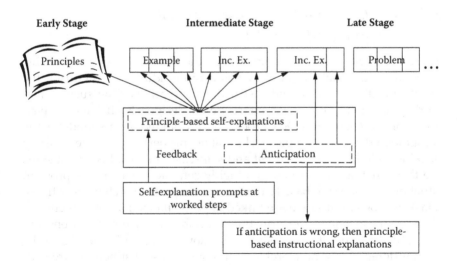

Figure 8.1 An instructional order model for example-based learning (Inc. Ex.: incomplete example). (From Renkl and Atkinson, 2007. In F.E. Ritter, J. Nerb, E. Lehtinen, and T.M. O'Shea (Eds.), *In Order to Learn: How the Sequence of Topics Influences Learning.* New York: Oxford University Press, 95–105. With permission.)

1. Introduce one or more domain principles.
2. Provide a series of worked-out examples. Do not initially provide instructional explanations while learners are studying the examples but rather encourage the learners to explain the examples to themselves (i.e., induce self-explanations).
3. At all steps that are worked out, provide prompts for principle-based explanations and provide feedback regarding the correctness of learners' self-explanations.
4. Systematically fade out the degree to which the steps in the worked-out examples have their solutions provided, thereby requiring learners to anticipate the solutions to these steps. Provide feedback regarding the correctness of their solutions. The fading of solution steps should culminate in an example where learners have to solve the problem completely on their own. This creates a smooth transition from studying worked examples to problem solving.
5. Use instructional explanations as a backup in cases of incorrect anticipations to guard against gaps in knowledge. Instructional explanations should be carefully designed according to the principles given earlier.
6. Resort to problem-solving practice, as opposed to reflective cognitive activities, to heighten domain skills.

8.5 The four-component instructional design (4C/ID) model

There are various instructional design theories and models available in the literature that can be very helpful in guiding or structuring the development of training and instructional programs. One such framework is the four-component instructional design (4C/ID) model (van Merriënboer, Clark, and De Croock, 2002). Underlying this model is the recognition that most complex tasks that people perform involve a variety of skills including both *recurrent* and *nonrecurrent* skills. Recurrent skills are those that are performed in a highly consistent way across problem situations and are rule-based in the sense that particular characteristics of a task or problem situation are linked to particular actions. Nonrecurrent skills are those that are performed in a variable way across situations and are guided by cognitive schemas that allow the same knowledge to be used differently in a new problem situation. These distinctions are very analogous to the distinctions that were made above between procedural and heuristic tasks.

The extent to which these recurrent and nonrecurrent skills are required varies considerably across tasks. Many simple tasks that do not require deep meaningful learning, such as learning to use an appliance, primarily emphasize recurrent skills. However, the distinction between recurrent and nonrecurrent skills is important because the learning processes associated with the formation of these skills and the instructional methods that support these processes are fundamentally different.

Generally, most instructional models, including the 4C/ID model, assert that for any instructional or training program to be effective three basic conditions must be met. First, training and instruction programs must promote the achievement of the following learning goals: the coordination of all skills required for a task; the integration of new skills with previously acquired knowledge and skills; the formation of cognitive schemas; and the ability to apply the new knowledge and skills across variable situations. Second, it must be recognized that people learn in different ways. Third, the cognitive limitations and capabilities of the learner must be considered.

With regard to the third condition, most current theories and models of instructional design maintain that the effectiveness of an instructional program will be enhanced if consideration is given to the learning process and the limitations of the human information-processing system. This is especially true when designing programs for older adults who often experience some degree of age-related changes in cognitive abilities (Chapters 2, 4, and 7). Meaningful learning usually occurs when new knowledge and skills are encoded into long-term memory (LTM) and can be recalled and applied at a later point in time. For this to occur, information must be attended to and processed by working memory (WM). The subsystems for processing verbal information (e.g., verbal instructions or printed text) and pictorial or spatial information (e.g., layouts of controls on devices) are limited in terms of the amount of information that they can process at a given point in time and the duration for which the information can be maintained. Therefore, in most learning situations the learner will need to make decisions about what information to attend to and select for further processing. The learner also needs to organize and integrate the new information, build mental models and schemas, and connect these models and schemas to existing knowledge and schemas in LTM.

Overall, the 4C/ID model is a framework that is intended to facilitate the learning process by enabling the development of mental models and schemas and strengthening the development of recurrent skills while also ensuring that the total cognitive load does not exceed memory resources. The four model components refer to learning tasks, supportive information, part-task practice, and just-in-time information. An overview of this model is illustrated in Figure 8.2.

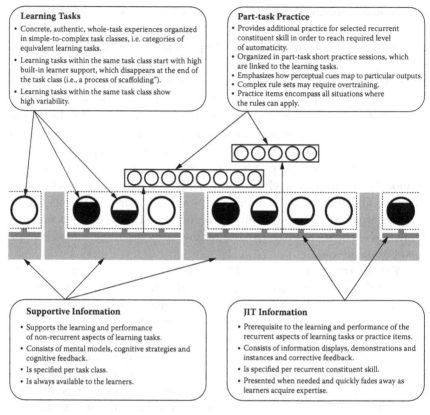

Figure 8.2 The four-component instructional design model. (Adapted from van Merriënboer et al., 2002. *Educational Technology, Research and Development,* 50: 39–64. With permission.)

8.5.1   Learning tasks

As implied in the various sequential elaboration strategies underlying elaboration theory (Section 8.3.3), current perspectives in training and instructional design emphasize *situated learning*. In this type of learning, the learner is immersed in realistic learning contexts and exposed to holistic and authentic learning tasks (concrete meaningful examples) that correspond to the actual tasks they will confront at home or at work. Exposure to meaningful learning tasks promotes the construction of cognitive schemas—complex connections of ideas and knowledge that become stored in LTM—which enables the learner to generalize knowledge to new situations beyond the training environment. In this regard, a well-designed set of learning tasks represents a core component of the 4C/ID training program.

Learning tasks should represent the entirety of the to-be-learned material and provide practice on both the recurrent and nonrecurrent aspects of a task. A key to the design of learning tasks is instilling variability through the provision of a wide variety of exemplars of a task. This necessitates identifying all of the critical elements of a task (e.g., through task analysis) and a wide variety of contexts in which the to-be-learned task might be performed. For example, when teaching someone to use an x-ray machine in a hospital's emergency room environment, examples should include the different constraints in positioning the equipment that the technician might confront due to the nature of the patient's injuries, and how to make decisions related to the need for retaking or repositioning x-rays given the quality of the x-ray images.

To create highly variable practice situations, learning tasks can vary in terms of the saliency of defining characteristics, the context in which the task has be to performed, the familiarity of the task, or any other dimensions that might exist in the environment where the task will be performed. Learning tasks should also be designed in a manner such that the learner is progressively weaned from support and is gradually required to assume more responsibility for performing the tasks. If possible, opportunities should also be provided for the learner to encounter task situations that are different from the examples demonstrated during training. This might involve having them complete unique tasks during "homework" or through transfer tasks.

The benefits of using the whole task approach when designing learning tasks are that it promotes coordination and integration of the constituent skills and provides the learner with a holistic vision of the to-be-learned material. However, as noted in the discussion of sequencing strategies, if the to-be-learned material is complex and has a high intrinsic cognitive load, using complex learning tasks at the beginning of a training program could yield excessive cognitive load, especially with older learners, with potentially negative effects on learning, performance, and motivation.

For such complex learning situations, a basic way to reduce the total cognitive load is through *scaffolding*. This involves sequencing the learning or practice tasks in a set of tasks so that high levels of support are provided early and progressively lowered toward the end of a particular set of practice tasks. In this way, the learner is progressively weaned from support and is gradually required to assume more responsibility for performing the tasks. For example, the initial tasks in a given set of practice tasks can be accompanied by worked examples, followed by partial worked examples (that require the learner to complete parts of the exercises), and finally tasks the learner needs to completely solve on his or her own. In the extreme case (i.e., full learner support), the learner can watch an expert perform and rationalize the actions taken when performing the whole task. In general, the scaffolding should be designed so that the challenge of the learning tasks is maintained, while never allowing the learning tasks to become too excessive in their cognitive demands.

There are several reasons why different sets (sometimes called "blocks") of practice tasks (Figure 8.2) may be incorporated into the instructional program. One reason is that such a scheme can easily accommodate the use of a simple-to-complex training sequence by enabling lower-level subskills in a simplified version of a whole task to be imparted early in instruction; these skills could then serve as prerequisites for the more complex holistic tasks to be performed later in the instructional sequence. The initial block of tasks may rely on easy-to-understand examples that can be presented using a scaffolding approach. Subsequent blocks of practice tasks could then use more difficult versions of the whole task, perhaps by embedding into the holistic practice tasks details that are more vague or larger in scope. A second reason for using blocks of practice tasks is to capture different categories of knowledge that need to be learned. The different sets or blocks of tasks could then address these different categories.

Another issue regarding the design of blocks of holistic practice tasks concerns how variable the practice tasks should be within a block. The issue of whether tasks with structurally similar contents should be taught successively, or whether the instruction should allow for alternation among tasks with dissimilar contents, has been a source of debate. Those who support use of highly variable instructional sequences would claim that such an approach leads to more flexible knowledge that would enable the learner to recognize when to apply specific skills and knowledge. The implication for the 4C/ID model is that by increasing the variability of the holistic practice tasks within a block (although presumably still maintaining the tasks within a block at about the same degree of complexity), one is promoting in the learner a greater ability to abstract and generalize to a wider class of problems. However, there is also a viewpoint that better learning outcomes can be achieved if concepts are grouped based on resemblances in their structural features, that is, by having less variability among tasks within a block.

Some of the discrepant findings in this area may be due to the nature of the learned material. If the tasks require instruction on large amounts of procedural skills, highly variable instructional sequences may be beneficial. However, if the emphasis is on the learning of concepts, a relatively invariant instructional sequence of examples or practice tasks may be beneficial, perhaps because such a strategy reinforces the concepts and minimizes confusion among the concepts.

The superiority of one approach over the other may depend on whether the learner has the cognitive ability to process highly variable instructional sequences, which has translated into the recommendation within the 4C/ID model of ensuring that the learner has attained the required skill level on a new skill before introducing greater variability into the instructional sequence. This recommendation is especially relevant to older learners who may be experiencing declines in cognitive abilities. Thus, although greater variability in the block of instructional tasks is likely to invoke more flexible cognitive schemas, it should be introduced with caution, especially with older learners, as they may be more prone to experiencing cognitive overload in response to highly variable instructional sequencing in practice tasks.

8.5.2 *Supportive information*

Supportive information refers to information relevant to the conceptual aspects of task activities. This type of information helps to promote a complete understanding of the task, in part by facilitating within the learner the construction of schemas, and thus enables the learner to gain nonrecurrent skills effectively. Supportive information can include the provision of a conceptual model of a task, with examples of how the various elements of the model are organized and interact, pertinent facts and rules, strategies for problem solving, and rules-of-thumb for handling certain aspects of problems. For example, with respect to performing a type of maintenance operation in a work setting, supportive information might include information related to identifying situations that warrant switching to a different tool and how to detect a dangerous situation involving energy hazards. This type of information should not be specific to a particular task within a block of tasks, but rather be pertinent to the entire group of holistic tasks constituting a block, as illustrated in Figure 8.2.

However, although in principle this information should always be available to the learner, for example, through flowcharts or a menu-driven system, for instruction on certain types of nonrecurrent skills this may be challenging. More direct interventions through face-to-face instruction or intelligent agents may be required to help the learner, and especially the older learner, construct mental models and cognitive strategies. In either case, to minimize the possibility for cognitive overload it is important that supportive information not become an intrusion while the learner is in the process of performing a learning task. Otherwise, the supportive information may become transformed into extraneous cognitive load as opposed to effective cognitive load.

8.5.3 *Part-task practice*

Part-task practice is required if the learning tasks do not provide sufficient practice on the recurrent aspects of tasks to achieve the desired level of automaticity. Essentially, part-task practice promotes the strengthening of IF–THEN rules and associated actions by providing the learner with practice on a particular recurrent skill. Part-task practice typically involves repetition, where learners repeatedly perform the recurrent skill. Examples include the synchronized motor activities required for performing a specialized type of welding operation or the series of actions needed to move and align information from one part of a database to another in a software application. In such cases, intensive overtraining may be required to make the skill fully automated.

The appropriate instructional design strategy is to encourage repetition of the procedural elements of the task through short periods that are spaced rather than long concentrated periods. The set of practice items should include examples of different situations where the rules apply and emphasize how certain perceptual cues map onto particular outputs. Part-task practice should also be linked to the whole task exemplars (learning tasks) used in the training program. For older learners, it is especially important to pay close attention to signals (explicit or less obvious) that indicate the need to go back and review a step-by-step procedure. Although this is harder to do in learning environments that are stand-alone such as online training platforms, even these learning environments should be sufficiently flexible to allow the learner to self-regulate her learning and easily activate the specific type of information or practice that is needed at a given point in time.

Part-task practice is especially important for older learners. To the extent that the recurrent skill is critical for performance of the overall task, the negative implications of the inability to grasp these skills could become amplified in older learners, as they may worry, to the point of becoming preoccupied or anxious, that they will not have the ability to meet task demands that entail use of these skills. These kinds of distractions can then hamper the older learner's ability to process other task-related information (e.g., that which is related to nonrecurrent skills).

8.5.4   Just-in-time information

Just-in-time (JIT) information is prerequisite information needed for various recurrent aspects of task performance that can provide learners with the knowledge they need to perform the *how-to* elements of a task. For example, it could be in the form of a demonstration of how to solve a particular type of problem with an accompanying explanation for the actions performed and the reasoning underlying those actions. It could also include feedback on the quality of one's task performance and explanations on the nature of the errors and necessary corrective actions at a level that allows the learner to understand how and why task performance was not correct.

Within a block of tasks, JIT information would be presented for each recurrent skill that could benefit from such support. The key is not to present such information at times (e.g., before a training session) when the learner would have to remember such information out of context and recall it when needed, but rather to present it when the learner can integrate it within the context of a particular task within a block of tasks. Information should be presented precisely when learners need it during practice, for example, when a highly situation-specific rule is needed that associates particular conditions or stimuli with particular actions. This helps to minimize extraneous cognitive load, which can occur if the learner needs to integrate information that is separated by time or to attend simultaneously to different sources of information (e.g., reading the *how to* information while simultaneously performing the learning task). For older adults, more repetition of the procedural aspects of the task is recommended.

8.6   *Active versus passive learning*

In recent years, the term *passive learning* has carried a negative connotation in discussions related to methods of instruction and generally has been viewed as less effective than instructional environments that emphasize *active learning*. In reality, the distinctions between these two styles of instruction are not always that sharp. Perhaps the easiest way to envision passive learning is the traditional classroom training environment, where large amounts of primarily declarative knowledge (facts, concepts, rules, etc.) are disseminated to learners who passively absorb the material through visual and auditory modes, and presumably do not get to apply this material until some later point in time. "Page-turner e-learning" environments, in which learners take in much of this same information in a similar way but online, represent a more recent manifestation of a passive learning environment. Active learning environments, in contrast, have people learning by doing rather than viewing. The 4C/ID model, through its emphasis on the performance of holistic blocks of practice tasks, essentially espouses an active learning environment. Active learning also encompasses environments that utilize games (Chapter 10), various forms of collaborative learning (e.g., watching someone demonstrate how to perform a task and then following that person's example), and immersive simulations (e.g., being placed in a simulated problem-solving scenario that entails identifying and correcting a fault in an industrial control system).

However, as argued by Clark and Mayer (2008, p. 5), "Physical activity does not equate to mental activity, and it is mental, not behavioral activity, that leads to learning." The implication is that whether the instructional environment is passive or active in its characterization, the key is to ensure that it promotes the types of cognitive processing activities that are congruent with the learning objectives. Can more passive viewing actually lead to as good or more effective learning than more active forms of learning? Using experimental studies as a basis for such a body of evidence, Clark and Mayer (2008) provided instructional recommendations related to this issue for a number of different learning contexts.

For example, consider the issue of passively viewing worked examples as compared to actively engaging in practice exercises. Would learning be more effective if a person, say following the learning of a spreadsheet software application, performed six practice problems as compared to three worked-out examples that demonstrated how to solve the problem and three practice problems? Apparently, there is evidence that a combination of worked-out examples and practice exercises is more beneficial to learning than just practice. The rationale for this effect is that engaging in practice exercises imposes considerably more cognitive load than studying worked-out examples; consequently, there are fewer cognitive processing resources available for fundamental cognitive processes—selecting, organizing, and integrating—that are essential for learning.

For older learners and trainees, this raises the issue of determining what constitutes an appropriate balance of worked-out examples and practice exercises. Over-reliance on worked-out examples presumes that the older adult is focusing sufficient attention on the concepts and methods that underlie problem or task solution, which may be difficult for some older learners. Providing some form of demonstration of a problem solution is critical for older learners as it frees cognitive resources that they can use to develop concepts and mental models. However, the transition to practice on holistic problems or tasks from viewed demonstrations should not be too long, as such a gap could interfere with the reliability or retrieval of the recently learned information from LTM. The knowledge domain or extensiveness of training may also dictate to some extent what strategy might work best for older adults. For example, if the topic is complex, it may be best to alternate in cycles between worked-out examples or demonstrations and practice problems as new concepts associated with the instructional material are introduced. Otherwise, generalizing to these new concepts from previously learned concepts might be difficult. This approach is, in fact, consistent with the 4C/ID model. Specifically, the different blocks of practice trials could represent different aspects of the problem, with the earlier practice tasks within the block more oriented toward demonstrations that systematically transition to practice problems with no support.

So-called discovery-based methods in learning (Section 8.8.2) emphasize approaches that facilitate discovery of new ideas or solutions to problems, and one way this can be accomplished is through doing. For example, in some learning contexts, learners may be asked to fill in a graphical template (e.g., what type of insurance claim would go into Category A), or devise their own decision-making aid (e.g., generate three different sets of rules for dealing with issues associated with processing mortgage applications). An alternative to asking learners to generate such organizing structures is for the instructors to present these devices to them. The evidence appears to point to including such materials to help learners organize and understand material rather than to have them generate such devices in the hope of inducing deeper learning. This principle is probably even more pertinent to older learners. Generally, the risk of having older adults sacrifice extensive cognitive effort in the hope that they may achieve a deeper level of understanding instead of providing them with the organizing structure is too great and, for most learning situations, unwarranted. This is especially true when one considers that with older adults there are also heightened issues of fatigue and frustration (Chapter 6). Thus, with older adults one needs to be more cautious in designing instruction that imposes an increased burden on mentally constructing or doing, especially if there isn't a strong assurance that such activities will translate into more effective learning.

Another way of promoting active learning is through collaborative learning environments. The defining characteristic in this type of learning is some form of interaction among a group of participants who are receiving instruction. The question of whether this instructional strategy may have benefits for older learners is very difficult to answer as it would likely depend not only on the nature of the learning topic but also on the makeup of the group. An older adult who felt intimidated by the knowledge or personalities of other group members may fail to process the discussion or possibly even become more confused about the material by virtue of the discussion. On the other hand, it is not hard to imagine a socially oriented older individual who is motivated by the instructional topic benefiting from a collaborative learning environment. Given the lack of evidence currently available about the benefits of collaborative learning environments as compared to traditional lecture environments, and especially with regard to any benefits to older learners, instructors should be very cautious about employing a collaborative type of active environment as a basis for learning by older adults.

Related to the issue of passive versus active learning is whether learning is more effective with still visuals as compared to animations. Intuitively, teaching how something works would seem to be accomplished best with a computerized animation display that dynamically captures the details of the material compared to a series of still visuals presented in print media. In fact, in four experiments involving four different learning topics (e.g., teaching how a car's brake system works), Mayer et al. (2005) found that the still visuals resulted in better learning than the animated presentations. These studies, however, emphasized understanding how something works rather than doing something. Indeed, there is evidence for the benefits of animated presentations when the emphasis is on how to perform a task. This should not be surprising, as procedural training requires clear identification of cues for action and decision making. This is especially true when there are spatial and temporal constraints associated with the required behavioral activities, and animation affords a more faithful appraisal of these cues. For example, an animation that cues the learner to important features or relationships among features, or incorporates easy-to-use and effective pause and replay functionality may in fact result in better learning than a series of static snapshots.

Still, some explanation is required for the results of the experiments that indicated the superiority of static animations as compared to computerized animations for learning. Clark and Mayer (2008) offered two explanations. First, consistent with human information-processing limitations and capabilities (Chapter 7), the sheer magnitude of information presented per unit time in an animated display may exceed WM capacity. The learner may be unable to select, organize, and integrate relevant information effectively (i.e., keep pace, by resorting to necessary focused and divided attention) by virtue of the rate of information being transmitted from the animation.

The second explanation focuses on the benefits of presenting static animations and suggests that viewing still visuals may induce the viewer to mentally animate the pictures and thus achieve a greater degree of psychological engagement than would occur when viewing (or trying to keep up with) a dynamic animation. This latter argument turns the tables, somewhat, on the passive versus active learning environment question, as it implies in this case that the static (less active) presentation may actually induce deeper learning by more actively invoking cognitive processing. Thus, if the instruction is directed at achieving understanding of the material (how does something work?) rather than executing a procedure (how do I do something?), the benefits of static animations may be due to two interrelated ideas: these presentations decrease the cognitive processing load on the learner, which, in turn, provides them with the capacity to select, organize, and integrate information from these static snapshots and thus engage in deeper learning.

This argument should be even more persuasive for older adults: computerized animated displays, as compared to their static animated counterparts, should be even more challenging for learners experiencing some degree of age-related cognitive declines. However, if the animated display is part of a multimedia instructional presentation (Chapter 10) that consists of modules or elements focused on how to do something (e.g., how to use the search box or negotiate a side directory on a website), it is probably still more advantageous to present this information in animated form, provided that the total amount and rate of information being transmitted is relatively low and the learner has available easily managed tools for pausing and replaying the instructional material. Otherwise, creative uses of animated stills may be especially useful in teaching older adults about a topic, as the challenge in instructing older adults is often on ways to induce deeper learning.

It should be noted that the animated stills need not be presented in print media; they could also be presented as computer displays. The advantage of print media is that the entire display world of materials is present simultaneously for viewing. In contrast, the computer-display environment constrains the presentation of animated stills into a series of snapshots. Although the computer can be used to circumvent some of these constraints, for example, by enabling static stills to be presented alongside one another, or to alter the rate of presentation in order to simulate to some extent a dynamic display but at a rate more compatible with one's cognitive processing capacity, these interface manipulations may require some technical facility.

The tendency to move toward active as opposed to passive learning environments is not likely to abate by virtue of experimental evidence that indicates the virtues, in a number of situations, of more passive viewing. Rather, the goal should be to reap the benefits of inducing active engagement during instruction in terms of its ability to facilitate relevant cognitive processing in the form of schemas and mental models, while at the same time not overloading the learner's cognitive capacity. In training individuals for jobs in the work sector, inducing behavioral engagement is critical for capturing the various physical and cognitive aspects that constitute the context of the work tasks. It is even more risky to expect older learners to generalize from more passive instructional settings to these work task contexts. At the same time and consistent with sequencing principles in instruction and methods such as the 4C/ID model, the active engagement process needs to be thought out, especially for older learners. Thus, for example, features of the active learning environment that produce extraneous cognitive load that is not directly related to the learning objectives should be eliminated as they can compete for needed cognitive processing resources. Likewise, practice exercises should be distributed within the learning lesson as this will lead to better retention, and it is essential that tailored feedback capable of providing the rationale for correct and incorrect responses be provided.

8.7 The teach-back and teach-to-goal strategies

There are a number of less formal instructional situations, such as when physicians provide instructions to their patients regarding protocols that need to be followed, for which comprehension of the instructions is critical. Concerns related to comprehending instructions also arise in research studies in which participants are required to consent to the research protocol but may not understand the documents they are reading or information about their participation in the study that is being told to them. Sometimes these problems stem from oral communication, for example, by a physician that is too advanced for the patient, or documents such as consent or privacy forms that are written at too advanced a level for the participant. There are many characteristics associated with the recipient that can undermine comprehension of instructions, including low educational achievement, older age, and low health literacy.

One relatively simple instructional technique that has been used to address this concern has been referred to as the *teach-back technique*. Essentially, this technique is intended to enhance communication and confirm understanding of instructions, whether provided face-to-face or in written form, by asking people to recall or explain in their own words what was discussed or written. For example, in a study examining methods for improving cardiovascular medication adherence of a predominantly low-income population being administered care in the primary care clinics of an inner-city teaching hospital, the interviewer assessed patients' comprehension of the written informed consent and HIPAA (Health Insurance Portability and Accountability Act) forms using teach-back techniques, and based on this feedback, provided additional instruction as needed (Kripalani et al., 2008). The authors of this study noted that this interactive process is presumed to enhance both short-term and long-term recall of pertinent information; thus, the technique should generally be expected to be particularly useful for older adults.

In this study, participants were randomly assigned to one of four groups: usual care, an illustrated medication schedule, refill reminder postcards, or both interventions for a period of one year. Teach-back was used as a basis for testing comprehension of the instructions. If the patient responded correctly, the interviewer proceeded to the next question. If the patient responded incorrectly, a teach-to-goal approach was used, whereby the information of interest was restated and the question was revisited within a few minutes to confirm comprehension again. The primary outcome variable used in this study was a measure of comprehension that indicated the extent to which patients were able to successfully teach-back, on their first attempt, eight items comprising consent and HIPAA questions. An example of a teach-back question pertaining to information related to the patients' understanding of the randomization process of the study group was, "Tell me about the four groups and what we'll give you if you're in each one of the groups. You can look at the picture if you want."

The investigators found that only age and health literacy (as measured by the Rapid Estimate of Adult Literacy in Medicine or REALM) were significant independent predictors of comprehension when other variables (such as years of education, race, gender, and employment status) were included in their statistical model. The importance of this study is that it used the teach-back method to validate comprehension of information and was able to determine that, despite recommended steps undertaken to simplify the information, concerns existed for older and low health literacy adults. Among the benefits of the teach-back method in such situations is that one can find out in real-time how well the person being informed or taught understands the material. This enables specific points to be clarified immediately, and thus for validation that the material has been made clearer by asking the individual to once again teach-back the material. In a teach-to-goal strategy, the material is repeated until understanding is achieved. With older people, one concern with the teach-to-goal strategy is that it may be perceived as a message that they are incompetent on some level, which could trigger anxiety concerning their self-efficacy (Chapter 6).

The concern for low health literacy has been considered a key quality and patient safety issue by the Institute of Medicine and other important health groups. Patients with low levels of functional health literacy are likely to have difficulty recalling and comprehending medical information. This is true, especially if they also have chronic health conditions such as diabetes mellitus, as these kinds of patients typically must contend with an array of challenging health management tasks that include coping with complex treatment regimens, managing visits to multiple physicians, and monitoring themselves for changes in health status. These concerns, and the recognition that educating the patient is considered one of the three main functions of the medical interview have been largely responsible for the permeation of the teach-back technique into the physician–patient interactive process, or what Schillinger et al. (2003) referred to as the *interactive communication loop*. For example, these investigators found that when physicians used this technique to assess recall or comprehension among diabetic patients, those patients were more likely to have adequate glycemic control than patients whose physicians did not assess recall or comprehension.

Ruiz et al. (2011, personal communication) described the use of avatars (Chapter 10), which are digital representations of people, in the development and application of *virtual teach-back* (vT-B). The vT-B technique, which can overcome limitations attributed to the traditional "live" teach-back method such as lack of standardization in its application and excessive demands on time that can potentially be imposed on already overly burdened staff, makes use of developments in computer science, gaming, multimedia e-learning, and communication research. It can be summarized as follows:

Step 1: The avatar, using a pre-recorded human voice, provides a brief explanation about the concept or action to the patient, communicating only one idea at a time and concluding the explanation by asking the patient a question to verify understanding.

Step 2: A menu of three to five text options appears that may include correct and incorrect items and an option that indicates lack of understanding.

Step 3: Using touch or a mouse, the patient verifies comprehension by choosing one option from the dialogue menu.

Step 4: Immediately following the user's selection the avatar provides elaborated feedback consisting of a verbal explanation. This explanation is dependent on the user's selection and may include one of the following: *correct understanding*, where positive verbal and non-verbal reinforcement is provided by the avatar, which then proceeds to continue with other explanations; *incorrect understanding*, where negative verbal and nonverbal feedback is provided by the avatar, which then proceeds to address the misunderstanding by clarifying the response with additional explanations and concludes by again asking the patient a question to verify comprehension; and *no understanding*, where the avatar provides a new explanation at a lower level (below sixth-grade reading level), and concludes by asking the patient a question to verify understanding.

Step 5: After the avatar completes the explanation in the case of an incorrect or no understanding response, a new menu of three to five options in text format appears that may include correct, incorrect, or no understanding items. Steps 3–5 are repeated until the concept or skill is demonstrated through selection of correct options (Figure 8.3).

The design of instructions to patients, especially older people or those with low literacy, can benefit from other types of interventions as well as teach-back strategies, and this has been an active area of research. For example, in addition to teach-back, where the patient listens and repeats the instructions, it is also important to use other senses, for example, preprinted forms that can be customized for each patient to remind them

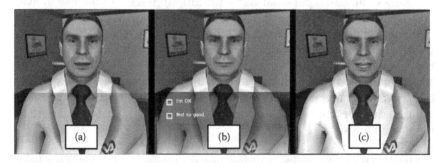

Figure 8.3 a) The avatar provides a verbal explanation; b) a menu of options appear for the patient user to choose from; c) the avatar physician provides verbal feedback. (From Ruiz et al., 2011, personal communication.)

about medications and side effects. Such an approach is consistent with the basic human factors design principle of incorporating redundancy in the presentation of information, where information presented in different sensory modes (e.g., auditory and visual) or different dimensions of a particular sensory mode (e.g., text and pictures) could help to reinforce the comprehension or recall of the information. For example, Morrow et al. (1998) demonstrated that pictorial illustrations combined with text significantly improved knowledge of dosing accuracy and timing as compared to text-only instructions for younger and older people, especially in situations in which a greater complexity of medication instructions was given, which is often the case for older people.

The benefit of pictorial illustrations in these instructional scenarios is that they may help draw attention to the written material. Of course, pictorial illustrations could actually hinder understanding if these forms of information are not designed or presented well, so that they actually generate more confusion than resolution of the written material. As always, pilot testing and field testing are necessary to ensure that the illustrations are compatible with the target population.

Likewise, in designing written materials for patients, such as brochures to accompany teach-back methods, proper human factors guidelines should be followed to ensure, especially for older adults, that this material will be easily perceptible and understood. Specifically, it would be advisable to avoid italics and abbreviations. Recommendations for effective written materials include: a great deal of white space between segments of information; eight- to ten-word sentences in an active voice; a conversational style incorporating relatively simple words; shorter paragraphs; and cues such as arrows, underlines, and bullets to guide the reader's eye to the most relevant information and bold type to emphasize this material (Mayeaux et al., 1996). For example, in comparing the following instructions (illustrated by Mayeaux et al., 1996, p. 210), clearly the second message will be better understood by more people:

- "Use of nasal saline lavage, followed by nose blowing, greatly decreases nasal congestion."
- "Spray the salt water into your nose and then blow your nose. Now you can breathe better."

8.8 Other ideas and approaches to instructional design

8.8.1 Anchored instruction

The method of *anchored instruction* is a technology-based framework for learning that was developed by the Cognition & Technology Group at

Vanderbilt (CTGV) under the leadership of John Bransford, to whom this method is generally attributed (CTGV, 1990). The key idea is to "anchor" learners to a problem context through which they investigate the problem, identify gaps in their knowledge related to the problem, review or study materials that can help them solve the problem, and finally develop solutions. The learner can be facilitated by a teacher who can coach the learner through the process, which could involve searching online for sources of information to explore and manipulate, including virtual simulations.

Anchored instruction has associated with it a number of principles. These include the use of a realistic type of task or event that forms the basis for the instructional experience, such as an industrial worker trying to control the formation of steel from raw materials or manage an inventory system. The learner plays the pivotal role, and thus takes "ownership" in this focused task or event, for example, by assuming the role of the worker and becoming actively engaged in trying to find a solution; an emphasis on the development of "deep" knowledge through the presentation of multiple scenarios, which can purportedly be transferred to other similar contexts; the presentation of the problem in a narrative format, presumably to enhance and reinforce the subjective experience of playing the role and the ability to create a story that is embedded with relevant data; and the promotion, through the available learning materials, of solutions by enabling the learner to identify subproblems and connect these subproblems in ways that lead to solutions to the problem.

Anchoring older learners could provide a useful way of motivating these individuals to explore the content of and pursue engagement with various learning materials, and thus develop the kind of deeper knowledge that may prove useful across different scenarios. For example, when training older adults in the work sector, anchored instructional techniques could be used to help establish a stronger context through which knowledge could be gathered and assimilated, thereby helping to promote problem solving related to the kinds of issues that are expected to be confronted on the job. This can be accomplished by designing the instruction to be anchored in a case study or problem in which the worker, whose role the older learner is assuming, is engaged. Not only may such approaches to training be more motivating to older adults, but they would also preclude these learners from having to retain large amounts of information in memory prior to solving the problem. A key challenge in adapting anchored instruction for older adults is to ensure an adequate collection of learning materials that can be easily understood and that can lead to knowledge relevant to the problem, and to offer, either through face-to-face or computer-based instruction, guidance regarding how to best pursue these materials. In addition, provision of feedback needs to be embedded in the anchored instructional method concerning where gaps in knowledge and in problem resolution still exist, as well as strategies that could help the learner close those gaps.

8.8.2   Discovery learning

Discovery learning, similar to anchored instruction, is an approach that was intended to introduce new and innovative ways for students to learn as compared with traditional classroom lectures. Instead of the more conventional "didactic" mode of teaching, the discovery learning paradigm, which is generally attributed to the ideas of Jerome Bruner (1966), encourages learners to ask questions, formulate tentative answers, and deduce principles from practical examples and experiences. By promoting guessing and hunches, it seeks to develop a mindset toward inquiry and hence an attitude that will enable learners to discover the principles or concepts that are to be learned.

As with anchored instruction, discovery learning supports active engagement of the learner and personalizes the learning experience. This can be intrinsically rewarding and thus highly motivating for the learner. It also builds on the learner's prior knowledge and understanding, and so may be suitable for those older adults with a rich foundation of lifelong knowledge and experiences. According to Bruner, this approach to learning can aid conserving memory as the emphasis is not on rote memorization but on finding solutions. This is not meant to imply that the emphasis on discovery should replace conventional methods of instruction, but rather that it become an additional goal of learning that can help in learning concepts and problem solving.

One possible advantage of discovery learning that has been pointed out is that posing a problem in a more goal-free way, as opposed to the more conventional way, such as, "Given the following conditions, find the velocity of Y if X changes by x units," results in less cognitive load. Most conventional accounts of problem-solving rely on Newell and Simon's (1972) influential model, or what has been termed "means–ends analysis." This perspective on problem solving consists of formal operations on symbols individuals have in their heads; these symbols actually represent objects and the properties and relationships among objects that exist in the world. Problem solving would then consist of transforming these symbolic structures, using rules or operators that typically produce a set of intermediate goals, until an initial state (the given conditions) is transformed into the final end or goal state. The sequence of operations by which this transformation is accomplished is the plan, which would represent the sequence of actions that would presumably be needed for solving the problem in the external world. This model of problem solving, in addition to accounting for how humans learn (Chapter 4), has also been successfully applied to many artificial technological systems such as expert systems, robot planners, and intelligent tutoring systems.

In contrast, a goal-free or exploratory problem-solving perspective (e.g., "Given the following conditions, find out all about Y after X changes by x units") results in less overload of working memory. According to some theorists, a consequence of this lesser cognitive load is the ability for the learner to develop better schemas. This would appear to make discovery learning very suitable as a method for training older adults in concept formation and problem solving.

However, it turns out that the benefits of goal-free problem exploration may not be so apparent and depend on various attributes of the problem. If there is a "small search space" associated with the problem, that is, a relatively small number of ways of conceptualizing solutions to the problem, the goal-free approach may be superior to the more conventional means–ends problem-solving method. However, when the search space associated with the problem becomes larger, the load on working memory can actually become overwhelming as there is a greater variety of solution options to consider. Also, the learner may already need to possess some relevant or useful schema prior to engaging in discovery or exploration learning. Otherwise, in the absence of previous exposure to the problem domain, practice with worked examples could actually be more effective than exploratory learning and even conventional (means–ends analysis) problem-solving practice.

8.8.3 *Situated cognition*

Both anchored instruction and discovery learning reflect, to some extent, a perspective on cognition as being "situated" (Brown, Collins, and Duguid, 1989). From the standpoint of training and instruction, this means that a fixed and unquestionable orientation about the learning situation prior to engaging in the task or set of behaviors is not presumed. Thus, instead of presenting facts and rules prior to performing a task, this initial state or presumption of rigidity is loosened by building in a process of questioning and inquiry right from the start. It is probably accurate to assert that many current approaches to instructional design embrace certain premises that promote a leaning toward situated cognition. Specifically, they do so by supporting the idea that the learner needs to adapt to the larger contexts that characterize holistic tasks, and in the process be prepared to reshape one's thoughts and consequently one's learning about the problem or task. The learner's cognitive appraisal and approach to the task or problem become, in effect, "situated" to the context that they encounter.

In the idealized perspective of situated cognition, the mutually interactive activities of the person with the task or environment can lead to dynamic mutual modification. For example, if a worker is designing a newsletter for an organization, that worker's perceptions of part of what was drafted may instigate a reappraisal of the goals of the project, perhaps based on a discussion that took place at a recent meeting. This, in turn, could lead to a different perspective on the project that becomes manifest as a different design.

Most learning situations, however, are sufficiently structured so that the "cognitive freedom" at the disposal of the learner or problem solver is still somewhat limited. Often, there are relatively clear criteria, especially in training, that need to be attained. Still, the essential characteristics of situated cognition help establish new sets of boundaries in human–task interaction that can, in many instructional situations, promote more effective learning. For example, with respect to older learners, rather than enforcing a fixed way of responding to inputs when learning a computer application, the individual's characteristic sensory and cognitive appraisals of the situation can be acknowledged and embedded in a positive way into the instructional setting. Depending on how the older learner experiences the new materials or information being presented, different paths or branch points can be taken that are more consistent with the cognitive appraisals of the learning by the individual. This type of instruction obviously is not suited to group learning but rather is geared toward acknowledging, especially among older adults who may possess a wider array of experiences, that each learner may be experiencing a holistic task in a different context, and thus may require dynamic adjustments to instruction that are more compatible with their cognitive appraisals.

8.9   Recommendations

- In choosing between *topical sequencing* (a topic or task is first taught to some target level of depth) and *spiral sequencing* (a series of interrelated tasks is first passed through at a basic level), topical sequencing would likely present less of a risk to disruption of learning by older adults. However, some adjustments in instruction are needed to capture interrelated aspects in a sequence of topics or tasks.
- Use *hierarchical task analysis* to establish the proper sequence of training or instruction by ordering the skills needed so that a skill is never taught until a prerequisite skill has been addressed, and then mapping these skills to their specific task elements. In using this approach, it is critical to have a clear understanding of the older adult's starting knowledge.
- If a task consists of a series of steps that must be executed in a prescribed order, use a *procedural task analysis* to identify the corresponding procedural prerequisites in the order of their performance. If the material corresponding to individual steps is large, consider

decomposing the instruction into manageable units across more sessions while preserving the emphasis on sequencing of steps.

- If the instruction is not targeting a number of small unrelated topics but topics relatively large in scope, present the learner with practice tasks that are holistic in nature in a progressive simple-to-complex sequence. Start with an initial task where the older learner has a reasonable opportunity to deal successfully with the task with some help.

- Consider segmenting the practice tasks into episodes or *blocks* of tasks. Ensure that the training program has a feature that always reinforces or summarizes the essence of the just completed block(s) of tasks and provides a logical lead-in to the upcoming learning episode to help the older learner associate needed knowledge with different scenarios.

- Avoid relying only on problem-solving exercises early on during training; instead present *worked examples* to emphasize important underlying principles.

- Try to induce self-explanations in the use of worked-out examples to promote reflection about the learning material, especially for older learners with low levels of prior domain knowledge. If providing self-explanations is difficult for these learners then consider, once they have acquired some knowledge, the presentation of worked examples with some blanks that the learner has to fill in so that the learner has to anticipate the next solution step.

- For complex learning situations, use *scaffolding* to reduce the total cognitive load. This involves sequencing the learning tasks in a block of tasks so that high levels of support are provided early (e.g., demonstrations or fully worked examples) and progressively lowered (e.g., partially worked examples) until the learner assumes full responsibility (e.g., through a problem-solving exercise). Never allow the learning tasks in a block to become too excessive in their cognitive demands.

- The transition to practice on holistic problems or tasks from viewed demonstrations or worked examples should not be too long; such a gap could interfere with the reliability or retrieval of the recently learned information from long-term memory.

- For complex learning tasks, the different blocks of practice trials should be designed to represent different aspects of the problem.

- Cautiously provide variability among tasks within a block, as this variability can help promote development of schemas related to the learning material in the memories of older learners. Too much variability, however, can induce confusion.

- Make it easy for the older learner to signal (to a trainer face-to-face or to a computer-based learning platform) at any given point during learning the desire for additional part-task practice on procedural task steps (i.e., recurrent skills).

- Avoid having older learners generate organizing structures as part of the training program in the hope of inducing within them deeper learning (e.g., asking them to generate rules for dealing with situations or to create templates). Instead, present this type of organizing information to them.
- Be very cautious about employing a collaborative type of active learning environment (i.e., where there is some form of interaction among a group of participants who are receiving instruction) as a basis for learning by older adults.
- Consider ways of using *animated stills* as compared to *dynamic animations* for learning situations that do not involve teaching how to do something. The use of animated stills can be less demanding of information-processing resources and may help induce deeper learning.
- For less formal instructional situations, such as when instructions to patients or workers concern protocols that need to be followed, consider using the teach-back strategy to ensure that the instructions are comprehensible to the older learner.
- If possible, *anchor* instruction in a case study or problem scenario that allows the older learner to assume the role of the person in the case study. This may provide increased motivation to learn and reduce the need to retain large amounts of information in memory prior to engagement on the learning task.
- Acknowledge the older learner's characteristic sensory and cognitive appraisals of the learning situation and try to embed these appraisals or tendencies in a positive way into the instructional setting.

Recommended reading

Beissner, K.L. and Reigeluth, C.M. (1994). A case study on course sequencing with multiple strands using the elaboration theory. *Performance Improvement Quarterly*, 7: 38–61,

Bielaczyc, K., Pirolli, P., and Brown, A.L. (1995). Training in self-explanation and self-regulation strategies: Investigating the effects of knowledge acquisition activities on problem solving. *Cognition and Instruction*, 13: 221–252.

Gagné, R.M. (1962). The acquisition of knowledge. *Psychological Review*, 69: 355–365.

Kalyuga, S., Ayres, P., Chandler, P., and Sweller, J. (2003). The expertise reversal effect. *Educational Psychologist*, 38: 23–32.

Scheite, K. and Gerjets, P. (2007). Making your own order: Order effects in system- and user-controlled settings for learning and problem solving. In F.E. Ritter, J. Nerb, E. Lehtinen, and T.M. O`Shea (Eds.), *In Order to Learn: How the Sequence of Topics Influences Learning*. New York: Oxford University Press, 195–212.

Shepherd, A. (2001), *Hierarchical Task Analysis*. London: Taylor & Francis.

Skinner, B.F. (1954). The science of learning and the art of teaching. *Harvard Educational Review*, 24: 86–97.

Sweller, J. and Cooper, G.A. (1985). The use of worked examples as a substitute for problem solving in learning algebra. *Cognition and Instruction*, 2: 59–89.

chapter nine

Instructional system design

9.1   Historical background

Instructional system design (ISD) refers to a systematic design method for developing a training or instructional program. This training program can, in principle, target large numbers of individuals, ranging from those employed in government and the military, to students in educational institutions, to workers in the industrial sector who require retraining.

Although the origin of the ISD label is obscure, the underlying concepts of ISD can be traced to the model that was developed for the U.S. Armed Forces in the mid-1970s. Specifically, as recounted by Branson (1978), the Center for Educational Technology at Florida State University worked with a branch of the U.S. Army to develop a model that evolved into the Interservice Procedures for Instructional Systems Development (IPISD), which was intended for the Army, Navy, Air Force, and Marine Corps.

The IPISD model encompassed five top-level processes: analyze, design, develop, implement, and control or ADDIC. These top-level headings or processes are very similar to those of the ADDIE model, which is discussed in Section 9.4. From a historical perspective, Molenda (2003) noted that the underlying concepts of the IPISD model can be found in an earlier handbook by Briggs (1970), whose model incorporates ideas similar to the IPISD model but without the ADDIC headings, leading Molenda to suggest that there are likely "many other tributaries leading to the main stream of ISD" (p. 35).

9.2   A human factors perspective to the ISD model

What are the benefits of using a systematic instructional design method such as ISD? Perhaps most important, it enables consideration to be given to many higher-level instructional issues. These could include the characteristics of the people who need to receive training, whether face-to-face versus computer-based training should be administered, what types of performance support aids might be required for the tasks that need to be performed, and what are the best ways to assess performance and to determine if the training goals have been met. Clearly, having a model or method directing the design of an instructional program is an efficient way of identifying important training issues as it provides a canvas upon which many considerations related to learning and instruction can be thought out and manipulated, including cost and feasibility concerns.

Because ISD represents a process or method for designing a training program, the program that is developed through the application of the ISD process can be considered to be a product or system. It then follows that the ISD process should conform to the same basic principles that govern the design of any product or system. In Fisk et al. (2009), we have elaborated on a human-factors-based product and system design process, especially as it applies to older adults. This process emphasizes a user-centered approach to design that iteratively involves the perceptions and assessments of potential users of the product through an array of usability testing methods, as well as the use of methods such as task analysis. When such human factors design considerations are incorporated into the generic ISD model, it can be described as a model containing four basic phases: a front-end analysis phase, a design and development phase, a full-scale development phase, and a final evaluation phase. The following is a brief description of these phases, especially as they would apply to training programs that might be implemented in industry, government, and military settings.

9.2.1   Front-end analysis phase

There are four types of analysis that are essential during the front-end analysis phase. The first is referred to as *organizational analysis*, which looks at the broad context of the job (e.g., organizational factors) to identify any factors that would bear on the need for, and the success of, a training program. For example, this analysis would be used to determine if, in fact, training is called for, given the nature of the tasks that workers are expected to perform and the levels of performance on these tasks that would be considered acceptable. A large hospital acquiring many new technologies may need to consider if a formal training program undertaken by the hospital's organization on the use of these technologies is needed. They may determine, for example, that it is more efficient to have the manufacturer of the technology train a few key personnel in the hospital, with these people then responsible, more informally, for training their colleagues. Similarly, the management of an organization that is planning to redesign many of its jobs may need to assess the relative benefits and costs associated with administering formal training on the new jobs as opposed to other options.

Assuming some form of training will be needed, *task analysis* (Chapters 7 and 8) is then performed to identify the knowledge, skills, and behaviors required for successful task performance. Of the many methods of task analysis, we have generally advocated hierarchical task analysis (Shepherd, 2001). Hierarchical formats, either graphical or tabular, can be especially useful because they reveal the inherent organization in the material. In particular, this type of representation is consistent with many of the methods of instruction that emphasize the nature of the sequence of material in learning, and thus can prove valuable for the more specific issues related to instructional design that would follow (Chapter 8).

Task analysis is followed by *trainee analysis*, which identifies prerequisite knowledge and skills trainees should have prior to beginning the training program, demographics (e.g., age, physical capabilities), and attitudes toward training methods. For example, if the objective is to design a training program for a redesigned job, there may be a *transfer of training* issue (Chapter 5); those people without prior experience with the job may require very different training from those who have performed the previous version of the job. Task analysis could, in fact, be used to determine if in this situation the possibility exists for *negative transfer* of training. If so, then training may have to be designed for those workers with previous experience on the original version of the job with the purpose of countering potential negative transfer effects, and designed for those workers without previous experience on this job with the purpose of promoting essential learning for the job.

These three analyses, organizational, task, and trainee, feed into a *training needs analysis*, which is ultimately used to determine the most appropriate approach to performance improvement. This could involve circumventing formal training and implementing the use of performance support, for example, if organizational analysis indicates that there is a large worker turnover rate for this job or that workers would be more motivated to perform this type of job with performance support. Organizational analysis might also reveal that the best approach to performance improvement is to redesign the job in a way that increases worker productivity without altering the basic knowledge or procedures the worker requires to perform the job. However, if task analysis indicated the need for workers to detect hazardous conditions during the course of task performance, or that difficult adjustments and procedures that can only be achieved through practice will be needed for the job, then a training program may be indicated.

Following these analyses, a set of *functional specifications* is written. These specifications include the training program's objectives, system performance requirements, and development constraints. These functional specifications mirror the *system specifications* linked to the front-end analysis activities associated with the general product design process. For example, the objectives might be to ensure that all nurses know how to detect malfunctions in a new type of monitoring equipment, how to link various configurations of displayed data to patients' medical states, and how to determine if the equipment requires resetting and the procedure for performing this operation.

The training system performance requirements should specify the means by which the program will accomplish the objectives. This requires specifying the instructional design strategies, which might include indicating where instruction on the development and use of mental models and analogies might be helpful or which aspects of the task may need to be trained to the point of being executed with little need for attention (Chapter 7). It also requires specifying human factors considerations such as indicating what the best interface might be for supporting demonstrations. Development constraints might include the availability of sufficiently knowledgeable trainers for demonstrations, or software developers to provide a way to interact virtually with the equipment from one's computer workstation. These specifications can then serve as a blueprint for the remainder of the training program design process.

9.2.2   Design and development phase

In this phase, the training system designer selects a method or combination of methods for instruction and proceeds with the design and development of the methods. By considering the information contained in the functional specifications, the designer generates a number of design concepts that would work for the problem. Cost–benefit analysis can be used to compare and evaluate the alternative methods or to select complementary methods that compensate for the disadvantages that may exist with some of the selected methods or approaches. These steps are generally iterated through many times, leading to the selection of a *prototype training design concept*. At this point a project plan can be formulated and documented, which includes budgetary, equipment, personnel, and timeline considerations. This plan may point to the need for a cost–benefit analysis to determine if the proposed training program is cost effective.

The prototype is then used for *formative evaluation* of the design concept. This would include usability testing (Lewis, 2006) on representative trainees to determine the usability, perceived effectiveness, and weaknesses of the training program's design through interviews, observations, questionnaires, and focus groups. Heuristic evaluations can also be conducted, whereby independent analysts assess positive and negative aspects of the design from the standpoint of an array of human factors considerations. These usability and heuristic evaluations can also be subject to an iterative process that steers the design into a more fully functional prototype.

9.2.3 Full-scale development phase

Following formative evaluation, full-scale development of the training program can proceed. Material is taken from the task analysis and translated into more concrete instructional steps using instructional design guidelines that might be assimilated from the various instructional ideas, principles, and guidelines from Chapters 8 and 10, as well as other chapters from this book and other books on instructional design. The focus should be on the acceptability of the training program to the trainees and whether the program appears to be meeting its objectives. In particular, the training program should be applied, if possible, to several naïve trainees who have not been part of the design process (in addition to other participants) in order to gauge the acceptability of the program through various usability testing methods applied in the previous design and development phase. The participants being tested should be evaluated on their knowledge and skill acquisition not only immediately following the training, but also after a period of time similar to the duration that would be expected to occur between training in the field and actual performance.

9.2.4 Program evaluation phase

This phase requires the specification of what criteria (i.e., variables) to measure (Chapter 11), who (which trainees) to use in measuring the criteria, and which contexts to use for evaluating the training program in the field. Pretest–posttest designs can be used, whereby knowledge and skills on tasks representative of those that will be performed on the job are measured on one group both before and after training. Another possibility is to use a control group design, for example, a design in which one group of randomly selected trainees receives the old training or none at all and the other group receives the training program to be evaluated. Finally, more distal organizational performance measures can also be evaluated, for example, the impact of the training program on the organization's productivity and performance levels.

9.3 How does age impact the human factors–influenced ISD model?

An obvious link between the ISD model and older adults lies in the front-end analysis phase, in which organizational analysis and trainee analysis are performed. One question an organization may need to consider is whether they want to recruit or include older people as trainees in the instructional program, or retrain older workers to perform new tasks. This is a potentially complex problem that requires addressing many issues, including the kinds of knowledge and skills demanded by the tasks and whether the knowledge and abilities of older adults are commensurate with the tasks' demands; whether older adults would have the motivation for performing these kinds of tasks; whether management considers older workers the right fit for these tasks, especially in the light of erroneous myths that they may embrace about the abilities and tendencies of older workers; and whether management believes the investments in training older workers will be economically justified (Czaja and Sharit, 2009).

If organizational and trainee analyses do deem that older trainees should be included in the instructional design program, it is essential that these older adults are adequately represented in the iterative usability testing cycles that are inherent to the design and development phase of ISD. Including older adults in usability testing helps to ensure increased variability in both subjective and task performance results, which provides designers with a broader range of potential problems and concerns that could be addressed in the iterative adjustments that typically need to be made to the training program.

9.4   The ADDIE model

Like the ISD model, ADDIE (analysis, design, development, implementation, and evaluation) represents a large-scale systematic framework for providing instruction. As discussed above, the historical roots of ISD have been somewhat difficult to establish. It turns out that tracing the roots of ADDIE may be even more challenging. When Molenda (2003) pursued answering what he considered to be one of the most frequently raised questions in instructional technology (IT) and instructional design (ID), "What is the original source for the ADDIE Model?" he discovered that mention of ADDIE did not appear in any of the existing dictionaries and encyclopedias of IT, in any of the published works on the histories of IT and ID, or in surveys of ID models. This led Molenda to conclude that:

> ...the ADDIE Model is merely a colloquial term used to describe a systematic approach to instructional development, virtually synonymous with instructional systems development (ISD). The label seems not to have a single author, but rather to have evolved informally through oral tradition. There is no original, fully elaborated model, just an umbrella term that refers to a family of models that share a common underlying structure. What everyone does agree on is that ADDIE is an acronym referring to the major processes that comprise the generic ISD process: analysis, design, development, implementation, and evaluation. (p. 35)

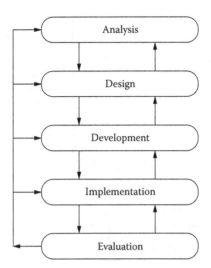

Figure 9.1 An ISD model depicting the sequential and iterative ADDIE processes.

It is generally accepted that when these processes are used in ISD models, they are not only sequential, but also iterative, as illustrated in Figure 9.1, but, as noted by Molenda (2003), "any claims about what the ADDIE Model *says* beyond this are individual inventions" (p. 35).

A relatively rare reference to the ADDIE model in the academic literature is found in Molenda, Pershing, and Reigeluth (1996) in their characterization of a systems approach to ID. They used the ADDIE concept as a model for illustrating the interconnections between the development of *instructional interventions* and the development of *performance improvement interventions*. The implication was that performance interventions, such as incentive programs, job redesigns, electronic performance support systems, and ergonomic redesign, should themselves be subjected to a process involving analysis, design, development, implementation, and evaluation to enable coherence between the instructional design strategy and the performance improvement intervention.

The website http://www.nwlink.com/~donclark/history_isd/addie. html, which we will refer to as the *nwlink Website* (2011), has presented a historical timeline related to the development and evolution of the ADDIE model. In nwlink's historical analysis of the origins of ADDIE, this model is noted to have first appeared in 1975 in the form of an ISD model devised by the Center for Educational Technology at Florida State University for the U.S. Armed Forces. Its basis was the realization of an increasing gap between the complexity of the military defense apparatus and the decreasing educational attainment of entry-level soldiers, with the potential solution to this problem being in the form of a *systems approach* to training.

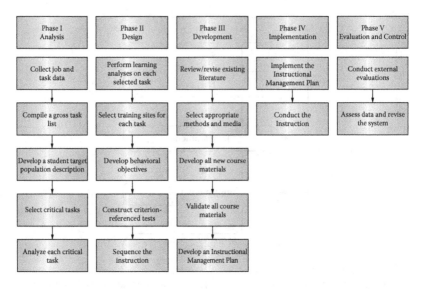

Figure 9.2 The five phases of ISD based on a presentation by Russell Watson in 1981. (From *nwlink Website*, 2011.)

When the ADDIE model first appeared as an ISD model in 1975 it was strictly a linear model. However, soon afterward it had taken on a more dynamic character. Apparently, even the U.S. Army, despite its characterization as a highly structured organization, could not design training in such a linearly constrained manner and by necessity needed to modify it into a more dynamic process. In 1981, Russell Watson presented a paper that contained a slightly modified version of the ADDIE model as developed by Florida State University in 1975 (*nwlink Website*, 2011). As shown in Figure 9.2, in his model the five basic phases were the same, but the steps within each phase were slightly modified.

Although it may seem obvious to some, it is important that the difference between ID (instructional design) models and ISD models, including ADDIE, be understood. ID models differ from ISD models in that ISD models have a broad scope and typically divide the instructional design process into the five phases depicted in Figure 9.1 (with possible modifications, for example, where the Design phase may be combined with the Development phase). ID models, in contrast, are less broad in nature and mostly focus on analysis and design; thus, they normally go into much more detail, especially in the design portion.

The broad scope and heuristic methodology that characterize ISD (and ADDIE by association) have often been criticized because learning designers are told what to do, but not *how* to do it. However, according to van Merriënboer (1997), it is this broad nature of ISD that gives it

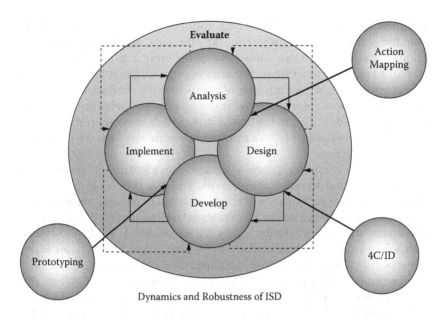

Dynamics and Robustness of ISD

Figure 9.3 The roles of instructional design methods, such as 4C/ID, and other instructional development approaches and tools, within the generic instructional system design model. (From *nwlink Website*, 2011.)

such great robustness. As for the how to do it issue, this would normally be handled by employing ID and learning models in conjunction with ISD models in an iterative and cyclical fashion. ISD is thus conceptualized as a "plug and play" model: essentially, you add other components to it on an as-needed-basis. For example, the ISD model depicted in Figure 9.3 has action mapping, 4C/ID, and prototyping plugged into it for designing a robust learning environment for training complex skills (*nwlink Website*, 2011). The 4C/ID model has already been discussed in detail (Chapter 8), and obviously as a plug-in ID model it can be replaced or combined with other instructional models. Brief overviews of the action mapping and prototyping plug-ins into the generic ISD model follow.

   Action mapping represents a method for involving learners in technical training in which the instructional designer can set up, often through the use of simulation, an example that is typical of the challenges that learners will face on the job, such as: "Bill has to go into the inventory control program to change the quantity of an order that hasn't shipped yet. He also needs to delay the shipping of another order." Using a scaffolding approach (Chapter 8), where the learner is initially given guidance that is gradually withdrawn, the learner can determine the appropriate menu items that need to be activated to complete the task.

Multimedia approaches (Chapter 10) could be incorporated to demonstrate the actions needed for particular problems, and to provide explanations for why certain problems require different actions. For example, the learner may see a fictional character appear and, through interaction with a menu, determine a number of needs that the customer would like to satisfy. Learners interacting with the simulation program may then be required to identify the best solutions, which may be in the form of "objects," and then select from a number of alternatives the best explanation for their choice.

The action mapping process focuses on identifying goals, what actions are needed to reach those goals, and the provision of activities that help learners practice each needed behavior. In general, the emphasis is on identifying the minimum information that the person would need in order to complete each activity. This very streamlined "cut-to-the-chase" approach to learning a task is likely to be highly suitable for older learners, especially when instantiated in a simulation environment, as the critical cues can be clearly perceived and identified in the form of objects that need to be acted on. Also, the nature of the required actions can be demonstrated in ways that diminish the possibility for confusion with other actions, all with the learner being subjected to minimal extraneous cognitive load (Chapter 7).

Prototyping refers to a general product or system design process that allows designers to examine and evaluate their concept, often in real-world usage, before committing to final design decisions. Obviously, in highly complex areas of design including ID, it could be very useful. In the case of ID, once the learning steps have been listed it is useful to take a sample of novices through the steps to determine if they can perform the task, or if modifications or additions to the learning steps are needed. Based on the results, adjustments to the design are made and this process is repeated until the design meets its objective, which is generally to enable learners to gain the necessary knowledge and skills in the most effective and efficient manner. This iterative approach, which was highlighted in Section 9.2, is often referred to as prototyping, where successive small-scale tests on variations of a limited function prototype are performed in order to permit continual design refinements. With this approach, the needs of the learners can gradually be better understood so that the design concepts and associated products can be translated into a unified whole product.

In reality, there could be hundreds of variations of the ADDIE model that could incorporate numerous approaches and tools as plug-ins to the more generic ISD process. Some of the human factors perspectives to ISD discussed in Section 9.2 represent such variations, as might any type of emphasis on the older adult, including those factors that address methods of instruction or evaluation of training. Because ISD-like models such as ADDIE are so encompassing, essentially providing a system-based template for designing instruction, all the possible details cannot be provided here.

In its timeline analysis of the evolution of the ADDIE model, the *nwlink Website* (2011), following its explication of van Merriënboer's (1997) perspectives of employing ID and learning models in conjunction with ISD models (Figure 9.3), goes on to briefly describe ADDIE in the 2000s as moving beyond a process model, working best with other performance models. This conceptualization is perhaps best captured by its depiction of ADDIE as an ISD process model that interfaces with a module that emphasizes various aspects and models of human performance (Figure 9.4). Within the *nwlink Website* (2011), users can learn more about any area of the model that they wish to by simply clicking on it. For example, clicking on "Doing" in the Learning Environment component will take the user to a webpage entitled "Learning by Doing" (http://www.nwlink.com/~donclark/learning/doing.html) that differentiates "doing" from "reflection," and provides a reference in this area (Clark, 2005) along with a site directory with links to a large number of topics relevant to learning (e.g., blended learning, learning styles, learning theories, and learner self-ratings).

Finally, another highly useful website that provides details concerning the ADDIE model is http://www.grayharriman.com/ADDIE.

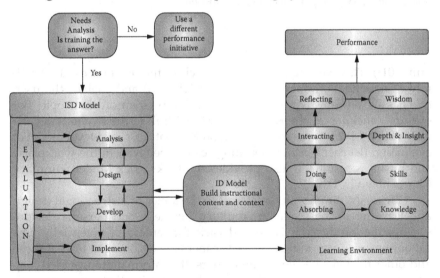

Figure 9.4 The ADDIE/ISD model, with an ID model plug-in to give the ADDIE model further design capabilities. Also illustrated is the link between the learning solution and the creation of the desired performance. (From *nwlink Website*, 2011.)

Table 9.1 Links to the Steps Within the Five Phases of the ADDIE Model

Analysis:	**Development:**
Needs assessment	Authoring
Audience analysis	Media creation/integration/
Content analysis	production
Technical analysis (course delivery and authoring tools)	Prototyping
Structural analysis (organization and course duration)	Processing
	Quality assurance
Design:	**Implementation:**
Identify goals	Promotion
Write learning objectives	Distribution
Identify entry behaviors	Reporting
Establish criterion reference	Maintenance
Research existing sites and resources	
Perform a content inventory	
Devise an instructional strategy	**Evaluation:**
Create maps and flowcharts	Perform evaluation
Design lessons and materials	(Kirkpatrick Model: Summative
Plan media utilization	Evaluation)
Design testing	
Design evaluation approach	
Design the interface	

Source: http://www.grayharriman.com/ADDIE.htm

htm (2011). This website breaks down each of the five phases of ADDIE into a sequential enumeration of steps (Table 9.1) and enables the user, by clicking on any step, to obtain more information on that topic. For example, clicking on the Kirkpatrick Model link in the Evaluation category (Table 9.1) provides information on Kirkpatrick's four-level model of evaluation (learner reaction, learning results, behavior in the workplace, and business results). It also provides the Kirkpatrick Model Summary Chart shown in Table 9.2.

Obviously, the enormous interest in the topics of instructional design and learning in the academic, military, business, and personal leisure settings has produced a large and varied inventory of potentially useful online resources, and interested readers are encouraged to visit these and other related sites. For some readers, the information provided may appear to be too cursory in nature, possibly requiring that the more traditional academic resources on these topics be pursued.

Table 9.2 The Kirkpatrick Model Summary Chart

Level	Evaluation Type	Description	Sample Tools
1	Reaction	How learners felt about the learning	Post-learning surveys
2	Learning Results	Increase in knowledge (before/after)	Pre- and posttest Interviews and observation
3	Behavior	Behavior on the job	Observation and interview
4	Business Results	Effect on business by learner	Management evaluation and business outcomes

Source: http://www.grayharriman.com/ADDIE.htm

9.5 Recommendations

- Consider using an instructional design model (ISD) such as ADDIE as a way of determining the overall needs of the instructional design (ID) program.
- If organizational and trainee analyses within instructional system design determine that older trainees should be included in the ISD program, these older adults should be adequately represented in the iterative usability testing cycles that are inherent to the design and development phase of ISD.
- In the design and development phase of ISD, select a method or combination of methods for instruction consistent with the principles of instruction for older adults, for example, as recommended in Chapters 8 and 10.
- In using action mapping to determine what actions are needed to reach a task goal, identify the minimum information that the person would need in order to complete the activity or task. This helps to ensure that the critical cues can be clearly perceived and identified in the form of objects that need to be acted on, and that the required actions can be demonstrated in ways that diminish the possibility for confusion with other actions.
- Iteratively employ ID models in conjunction with an ISD model so that the specific methods or approaches to instruction and training are consistently evaluated within the broader framework of ISD objectives, including job performance objectives.

Recommended reading

Gordon, S.E. (1994). *Systematic Training Program Design: Maximizing Effectiveness and Minimizing Liability.* Englewood Cliffs, NJ: Prentice-Hall.

chapter ten

Multimedia and e-learning

10.1 Multimedia

Many different instructional media or platforms exist. Examples of these include printed materials such as flowcharts, drawings, graphs, checklists, and manuals; physical mock-ups that can be physically manipulated; CDs and DVDs accompanied by sound, voice narration, animations, static snapshots, and text; synchronous (live) and asynchronous (independent, self-paced) *e-learning* platforms (Section 10.2); and various forms of simulations that can include role-playing as well as high-fidelity physical simulators. Multimedia systems represent a type of medium in which different types of information are integrated. Most typically, multimedia instructional systems integrate visual and auditory information. The visual information can be in the form of text or in a pictorial form presented as an animation or a series of static illustrations, and the auditory information can be in the form of spoken words, music, or other sound information. The tactile modality can also be an important component of a multimedia instructional system, particularly those that emphasize manipulation of various input devices or other objects.

Within the instructional design community, there are a number of those who believe that the type of instructional medium should not be the critical consideration in training and instruction. Instead, the emphasis should be on whether the instructional method promotes the kinds of cognitive processes needed for learning. In principle, all media are capable of imparting either effective or ineffective instruction. Ultimately, success will largely depend on whether and how consideration is given to the goals of the learning program, the nature of the learning tasks, the characteristics of the trainee population, and many other factors associated with how instruction is delivered.

However, there is some evidence to suggest that specific media work best for particular types of training situations. For example, watching a video may provide a better understanding of which buttons to press in using a product, whereas watching an actual demonstration of a maintenance task may make it easier to visualize how a tool should be held or positioned. Where change or movement over time is critical for understanding how something functions, the use of animation (e.g., in depicting how the values on a device change) may be more effective than static illustrations or photographs. Similarly, immersive virtual reality environments (VREs) such as flight simulators can promote learning by providing psychomotor feedback during practice by students. The unique benefits of certain high-technology learning environments have, in fact, led Moreno (2006) to conclude that the evidence supports a "media-enables-method hypothesis" due to "their potential to afford a variety of effective instructional methods" (p. 63). The key, according to Moreno, is to determine what types of learning methods can be afforded by the high-technology learning platform and, most important, the ways in which these methods can affect the human's cognitive processing so that effective learning is enabled.

With the increasingly expanding capabilities of computer-based instructional technologies, various forms of multimedia environments will continue to pervade the instructional arena. Clearly, these environments provide a wealth of creative approaches for administering training and instruction. Information can be presented in unique ways and across multiple sensory modalities that, if designed appropriately, can presumably maximize the human's information-processing capabilities. In addition, well-designed multimedia systems could afford a more pleasant, exciting, or motivating learning experience, which could keep the learner engaged and focused and thus produce greater learning effects.

Undoubtedly, multimedia systems have enormous information transmission capabilities. However, whether these capabilities derive from the multiple sensory systems that they can accommodate, the various controls of speed of information transmission, the interactive controls available to the user for manipulating the interactivity of the instructional system, or from the various levels of richness or resolution in which the sources of information can be presented, they need to be appropriately tethered to the human's cognitive processing limitations and capabilities. In this regard, the human's information-processing characteristics (Chapter 7), cognitive load theory (Chapters 5 and 8), and fundamental principles of learning (Chapter 4) are all critical considerations. Ultimately, it comes down to how well the learner can effectively process and integrate the various sources of information, either simultaneously or successively.

10.1.1   *The multimedia effect*

The principle that is perhaps most fundamental to multimedia instruction is that learning can be improved by including words (either printed text or spoken text) with graphics (charts, graphs, photos, static illustrations, videos, or animations). This principle is so basic that it has been referred to as "the multimedia effect" (Clark and Mayer, 2008). From an information-processing perspective (Chapter 7), the benefits from mentally representing learning material in words and in pictures can be summarized as follows: (1) we can make more effective use of our limited working memory (WM) capacities because words and pictures are processed by different WM structures; and (2) the opportunity exists for connecting the verbal and pictorial representations produced by the verbal and pictorial information, which can encourage deeper thinking about the material and thus promote more effective learning, especially in people with low knowledge of a subject domain.

Most conventional learning relies on words, printed or spoken. Although incorporating graphics can change the learning equation, how graphics can actually support learning may not be so obvious without an understanding of the different functions that graphics can serve. Table 10.1 presents a classification of different graphics functions. Generally, deeper learning can be promoted by incorporating graphics that help the learner understand the material (transformational and interpretive graphics) or organize the material (organizational graphics). For example, words (presumably printed text) in a multimedia presentation used to describe a quantitative relationship would benefit from the accompaniment of relational graphics, such as a line graph, whereas text describing changes over time can benefit from a transformational graphic. Table 10.1 also indicates how these different graphic functional types benefit the learning of five different types of content that can comprise learning situations: facts, concepts, processes, procedures, and principles.

The multimedia effect should, in principle, benefit older learners as it can make phenomena that are difficult to visualize or imagine visible, as well as show relationships that otherwise would be difficult to contemplate. For older learners, allocating limited WM capacity to trying to imagine how something works or how things are related to one another could take away cognitive processing resources needed for fundamental aspects of learning, including ensuring that basic facts and concepts are being processed. However, as discussed at length below and in Chapter 7, there could be many reasons why multimedia does not improve learning for older adults, including the fact that the executive control mechanism in WM may be subject to increased capacity limitations with aging, which could reduce the capability for combining verbal and pictorial representations in WM. It's possible that certain graphic functional types are less disruptive than others in terms of using cognitive processing resources, and research may be needed to identify the types of functions served by graphics that best support older adult learning in multimedia instructional settings.

In addition to the fundamental principle of combining words with graphics, a number of other basic principles or guidelines exist related to the design of multimedia learning systems. These principles include the alignment, in space and time, of words and corresponding graphics; how to apply narration in place of onscreen text; and how to use narration or

Table 10.1 Functional Categories of Graphics

Graphic Function	Description of Function	Examples	Type of Learning Content Supported
Decorative	Graphics that serve decorative or aesthetic purposes, or are intended to add humor	Photos of galaxies in a lesson on the origin of the solar system	These graphics do not enhance the message of the learning material
Representational	Graphics that illustrate the appearance of an object	A drawing of a tool for a maintenance operation or a screen capture for teaching Excel	These graphics support the learning of facts (e.g., a screen capture of a spreadsheet) and concepts (e.g., using diagrams to illustrate the structure of a relational database)
Organizational	Graphics that show qualitative relationships among the contents of the learning material	Matrices, maps, tree diagrams	These graphics support the learning of facts (e.g., a chart listing the names of components in inventory) and concepts (e.g., a tree diagram illustrating the relationships among personnel in a company)
Relational	Graphics that summarize quantitative relationships	Line graphs, bar graphs, pie charts, 3-D contour maps	These graphics support the learning of concepts (e.g., the relationship between a worker's energy expenditure and heart rate) and processes (e.g., how force affects the degree of valve opening)

(continued)

Table 10.1 Functional Categories of Graphics (continued)

Graphic Function	Description of Function	Examples	Type of Learning Content Supported
Transformational	Graphics that illustrate changes in time or over space	A time-lapse animation of the healing of a bone or an animated demonstration of a maintenance procedure	These graphics support the learning of processes (e.g., an animation showing how a virus invades a cell), procedures (e.g., a diagram indicating how to install a lighting fixture), and principles (e.g., a video demonstrating effective leadership strategies)
Interpretive	Graphics that make intangible phenomena visible and concrete	Depictions of Internet data transmission or drawings of molecular structures	These graphics support the learning of concepts (e.g., still diagrams to illustrate how a bicycle pump works) and principles (e.g., an animation showing how genes pass from parents to offspring)

Source: Adapted from Clark and Mayer, 2008. *E-Learning and the Science of Instruction: Proven Guidelines for Consumers and Designers of Multimedia Learning*, 2nd ed. San Francisco: John Wiley & Sons. With permission.

audio in combination with text and graphics. These principles, as well as others are discussed to a greater or lesser extent in some of the following sections, and are summarized in Table 10.2, often within the context of older learners.

10.1.2   Static media versus narrated animations

To illustrate the potential role of cognitive processing in determining the effectiveness of multimedia instructional systems, consider the task of instruction on how various things work (i.e., a process type of learning content). This issue was discussed in Section 8.6 entitled, "Active versus Passive Learning," in Chapter 8. Here, the focus is on the nature of the multimedia instruction and on how common beliefs about preferred modes of this kind of instruction may not be correct.

Table 10.2 Ten Guidelines and Associated Recommendations for Designing Multimedia Instructional Programs, with Emphasis on Older Adult Users

Guidelines	Recommendations
1. Do not overload the visual sensory channel. For example, presenting an animation of how the blood flows through the body, bottlenecks in blood flow, and effects of these bottlenecks on blood pressure, with text that appears below the animation, could overload the visual channel. It causes the user to "split" visual attention between the animation and screen text, which is particularly problematic for older people.	Consider using narration to present the text. Essentially, this serves to expand working memory (WM) capacity by making use of both the visual and auditory stores within WM. However, to ensure that the combined visual and auditory information does not exceed the older person's cognitive capacity, the presentation should be designed so that the speed of the narration is slow to moderate and that there is not too much information presented in the auditory channel. Also, mechanisms for repeating parts of the presentation should be clear and easy to manipulate.
2. Do not overload both the visual and auditory sensory channels simultaneously. For example, if the topic requires that the animation be relatively rich in detail, which may impose a large amount of information to be narrated as well, the learner, and especially the older learner, may not be able to process all of the needed information or may become distracted.	If the material has logical breaking points, segment the presentation. It may also be a good idea to have a lead-in to the next segment that acts as a refresher for the previous segment(s) and maintains the continuity between segments.

If segmenting the material is difficult, consider providing supportive information on terminology or other basic concepts prior to the actual training presentation. For the older learner, this supportive information would need to be well-rehearsed to ensure that it consumes very little cognitive processing when the learner encounters this material within the context of the larger presentation. |

3. Avoid presenting extraneous information. Instead, the focus should be on designing materials that are essential for developing the knowledge and skills associated with the learning objectives. In the example above, if one were interested in demonstrating how a damaged blood vessel can impede blood flow, do so after the basic learning material has been presented, not at the same time. Embellished narrations that focus on interesting angles associated with the topic can follow the basic learning material.

Similarly, avoid background music or excessive sound effects or other types of information that could cause the learner to have attention diverted, even if it is to a small degree, if this information is extraneous to the points being emphasized in the training or instruction. The same holds true for extraneous graphics, which can serve to distract the learner (by taking limited attention capacity away from relevant material) and disrupt the process of linking, in memory, relevant information from the learning materials. Adding sounds and graphics that presumably are interesting may seem like an appropriate design strategy as it may emotionally arouse learners which, in turn, may induce increased motivation and thus greater engagement with the lessons. Although adding extraneous sounds or graphics, as added text can be used than adding extraneous text may seem more innocuous to expand the key points or provide technical details that go beyond the key ideas, for the same reasons it can also be harmful to learning. In all these cases, the peripheral material can interfere, due to limited cognitive capacity, with the ability to process incoming material. With older learners in particular, this additional processing of information, through distraction, or disruption of needed processing of the learning content, however slight, could be critical given the possibility that these learners have diminished cognitive processing capacity.

To promote within the learner the ability to identify and process information essential for learning, cue the learner about how to select and organize material by stressing certain phrases in speech, adding headings, or highlighting certain images with arrows. This strategy should be used cautiously because if it is done too excessively it can lead to distraction, to which older adults are more susceptible.

(continued)

Table 10.2 Ten Guidelines and Associated Recommendations for Designing Multimedia Instructional Programs, with Emphasis on Older Adult Users (continued)

Guidelines	Recommendations
4. Avoid presentations in which the pictorial part of the material, whether it be in the form of an animation, graphics, or pictures, is distinctly separated from the words or text that are presented concurrently as an accompaniment to the pictorial material. An example would be to present an animation of what happens to blood that is attempting to circulate within the heart in one window and presenting text that explains what is happening in another window. By needing to invest additional attention to process which words go with which pictorial elements, often by back-and-forth scanning and sampling of textual and pictorial elements, the older learner in particular could be diverting limited cognitive resources to essentially an overhead activity that arises because of the way material is presented. Classic violations of this principle include the separation of graphics from the corresponding printed text due to scrolling screens, using links that lead the learner to a new window that covers information related to that in the window which was the source of the link, and placing text associated with a graphic on the bottom of a screen, including legends at the bottom of a screen that contain the names for numbered graphical elements displayed in the upper portion of the screen.	By integrating the presentation so that critical text is placed within the graphic, animation, or picture, the overhead associated with back-and-forth sampling, scanning, and matching of text and pictorial information is reduced or eliminated. Placing text near the pictorial information creates spatial contiguity. If there is insufficient space to place the text within the picture, do not attempt to clutter the pictorial elements as older adults generally have greater difficulty processing target information embedded in noise. Instead, consider using narration (see above) concurrently with the pictorial presentation to capture this additional textual information. Another possibility is to provide the learner with the opportunity to read this additional material in a separate window, but to do so after the integrated presentation has been processed. This way, the learner can decide if additional reading is warranted, and if so, to pursue this material while the integrated presentation is still available for consultation.

5. Avoid presentations that simultaneously present animation or pictures, narration, and onscreen text, where the text and the narration are redundant. Although some designers are taught to provide redundancy when presenting information as it reinforces the processing of information, in this situation the redundancy in narrated and onscreen text may be counterproductive, especially for older adults. This is because the learner may be devoting limited cognitive resources to ensuring consistency and resolving possible inconsistencies between the redundant narrated and onscreen textual materials. Redundancy is usually beneficial when there is no such processing required.

6. Avoid successive presentations of associated learning materials. For example, animations that rely on prior presentation of verbally narrated information require holding the latter information in WM and then using WM during the animation to select, organize, and integrate the information. This requires additional cognitive capacity and is thus especially difficult for older adults. Instead, ensure that the associated graphics are being depicted at the same time that the spoken words are presented. This creates temporal contiguity and thus better connections in WM between these materials.

Eliminate redundant narration and onscreen text. Because the visual channel is already absorbed in processing the animation or pictures, use narration alone for presenting the accompanying words. However, in the case where the presentation does not involve animation or pictures, learning is more likely to be improved if the presentation consists of concurrent narration and onscreen text (i.e., verbal redundancy) rather than narration only or onscreen text only, because adding either onscreen text or narration does not add to visual processing.

Synchronize the presentation of corresponding visual and auditory information. For example, if the graphic is an animation depicting the steps that a maintenance worker needs to take to repair a piece of equipment, the narration describing a particular step should be presented at the same time that the procedural step is illustrated on the screen. Likewise, narration should coincide with actions shown on videos.

If synchronization cannot be accomplished, for older adults consider presenting very small segments that alternate between narration and the corresponding animation information as this strategy reduces the possibility for overload. Temporal contiguity is critical for older adults as, with declines in the rate at which information is activated and processed, mutually dependent elements of information critical for learning may not get activated quickly enough. Thus, one element may have already dissipated by the time another becomes sufficiently activated, hindering the coordination and integration of information needed for learning.

(continued)

Table 10.2 Ten Guidelines and Associated Recommendations for Designing Multimedia Instructional Programs, with Emphasis on Older Adult Users (continued)

Guidelines	Recommendations
7. Adding relevant graphics or pictures to words results in better learning than just relying on words alone. This represents what was referred to as the "multimedia effect." Better learning is more likely to occur because the learner is induced to make connections between the words and the graphics, which results in a better representation of the materials in working memory, and ultimately in long-term memory. Also, without the benefit of accompanying graphics, older learners' cognitive processing resources can become consumed with attempting to conceptualize and reason about various processes or ideas, which can lead to frustration and also make them more likely to neglect important learning content.	As noted above, when presenting words and pictures, try to integrate them as much as possible to create spatial contiguity. Illustrations or drawings are sometimes better than actual pictures (e.g., photographs) as they provide better cues and less clutter, which can aid older learners in their selective attention processes. Thus, illustrations that are overly embellished to make them look more realistic may be counterproductive. Simplifying the graphics so that they contain the features essential to the learning topic may free up cognitive capacity in older learners that they could apply toward more meaningful or deeper processing of the material. Consider using actual pictures or a real video when they are deemed essential to the particular learning experience.
8. Have the learner manipulate instructional material (for instance, moving a cursor) if that is an important part of the learning experience, rather than passively observing how the materials should be manipulated. In addition, ensure that the computer system on which the multimedia presentation is being implemented is comfortable to interact with, attractive, and pleasant to use. These factors could help motivate older adults who may otherwise be wary of technology, and provide additional enthusiasm concerning the learning experience.	Use narration to instruct the learner on the required actions and provide feedback if the actions are not adequate. For older adults, ensure that critical interactions with input devices and other auxiliary technologies or objects are mastered prior to initiating the learning sessions so that cognitive processing resources are not allocated to the use of any of these devices during the learning sessions. Ensure that the visual display is large, that text is large and pictorial information is clear, and that onscreen windows are identifiable and clearly distinguishable. Older people generally need more illumination and contrast (which can compensate for reduced visual acuity) than younger people when performing tasks requiring reading or other visually demanding activities (contrast sensitivity is the ability of the eye to perceive a small difference in luminance). Because contrast sensitivity declines when the stimulus is moving relative to the viewer, dynamic displays such as animations or videos will be more difficult for older people to extract information from as they will need more time for their eyes to be fixated at a particular location to acquire the needed information.

Improving contrast is especially significant for luminous displays such as computer monitors, so ensure that there is adequate contrast between the screen objects and the screen background. However, with these devices increasing the amount of background illumination will actually make it harder to see the displayed information because ambient light (i.e., light from the environment) reflecting off the surface of the luminous display reduces the contrast between the displayed characters and their background. Therefore, increased illumination needs to be applied carefully in order not to exacerbate this problem. To mitigate problems with reflections (which can also cause problems with glare, to which older people are more sensitive), change the orientation of the computer screen to reduce reflections from the display surface and keep direct sources of light off the display surfaces.

Make certain the older learner has a comfortable chair, that the volume of any narration or sounds is adequate, and that controls for manipulating the presentation are easy to use. Provide for adjustability in font size and loudness.

For example, instead of a passive voice approach to instruction on how to perform a maintenance operation ("Extreme environmental conditions may erode insulation around wires and piping. Carefully inspect piping and wiring systems for signs of erosion."), use a friendlier second-person approach with more informal language ("It's not always easy to see wires and pipes that are in bad condition, so here are some good ways to make sure you don't miss damaged wires and pipes."). Also, use more polite speech ("You may want to move to the next screen.") as opposed to a more abrupt tone ("Click the enter key."). However, be careful not to overdo the degree of personalization as this may become perceptually salient to the learner and thereby distracting.

9. Personalize the learning experience by using a conversational approach that more closely resembles human-to-human conversation rather than a formal style that may appear more professional but is more likely to isolate the older learner. Creating a sense of partnership with the instructional system that is striking a more conversational tone with the learner is likely to create a sense of social presence, which may induce in the learner deeper cognitive processing. A more formal tone that simply focuses on delivering information may be less likely to engage the learner with the learning agent. Older learners may be more receptive to the impression of a social presence and benefit even more from such personalization.

(continued)

Table 10.2 Ten Guidelines and Associated Recommendations for Designing Multimedia Instructional Programs, with Emphasis on Older Adult Users (continued)

Guidelines	Recommendations
	To enhance the personalization experience, consider using onscreen characters who interact with the learner; when these characters assume the role of a coach that helps guide the learning process they are often referred to as pedagogical onscreen agents. These agents, which can assume various forms (e.g., a video of a talking head, a cartoonlike character, or a virtual reality avatar) can be used in various ways such as providing hints in practice exercises, explaining how the human heart works, or demonstrating a work procedure. The actual look of the onscreen agent is less important than the voice and personalization style of the agent. Synthetic voices or machinelike articulations should be avoided; not only are they less effective at inducing a social presence but it takes more mental resources to process such spoken words, which could lead to distraction and less effective learning by older adults.
10. Provide the opportunity to reflect on the presented material. This promotes better learning by allowing the materials to become better organized and for new connections to be established between the presented material and prior knowledge.	Provide opportunities for the learner to pause the presentation and to easily continue the presentation following the pause. Following a pause, the learner should be able to identify previous sections of the presentation easily, go back to any of these sections, and go to any point forward as well. Ease of and flexibility in navigation are essential for optimizing the benefits of reflection.

Table 10.3 Cognitive Processes in Learning with Static Illustrations and Text versus with Animation and Narration

Static Illustrations and Text Help Learners
Manage intrinsic processing because learners can control the pace and order of presentation (i.e., learner control effect).
Reduce extraneous processing because learners see only frames that distinguish each major step (i.e., signaling effect).
Engage in germane processing because learners are encouraged to explain the changes from one frame to the next (active processing effect).
Animation and Narration Help Learners
Reduce extraneous processing because animation requires less effort to create mental pictorial representation (i.e., effort effect), narration requires less effort to create mental verbal representation (i.e., effort effect), and computer control requires less effort to make choices during learning (i.e., effort effect).
Engage in germane processing because narrated animation creates interest that motivates learners to exert more effort (i.e., interest effect).

Source: Mayer et al., 2005. *Journal of Experimental Psychology: Applied*, 11: 256–265.

Although some kind of graphics are generally recognized as essential to such instruction, it is often assumed that the preferred mode of learning is through some form of animation (and accompanying narration). The logic is that this approach would provide a more realistic appraisal of the functionality of the system or the to-be-explained process, in contrast to an approach based on a series of paper-based static diagrams and accompanying text. In addition, and possibly because this dynamic type of presentation is more similar to the real process that it represents, it may be more motivating as well.

There is, however, another perspective to this issue. Specifically, the *static-media hypothesis* implies that the static media approach to information presentation will lead to better learning (e.g., based on tests of retention and the ability to apply what was learned to new problems) because of its ability to reduce extraneous cognitive load. This, in turn, can induce cognitive processing directed at aspects of the presentation that foster learning and promote what Sweller (2005) calls *germane cognitive processing* or what Mayer and Moreno (2003) refer to as *essential cognitive processing*. In either case, deeper processing of the relevant material can be promoted through mental organizational and integration processes. Support for the static-media hypothesis has important implications for training older adults, as these individuals are more likely to be susceptible to factors that could promote extraneous cognitive load. Some of the cognitive processing advantages afforded by the static-media presentation are summarized in Table 10.3.

Mayer et al. (2005) tested this hypothesis with college students who were given instruction on four different topics: the process of lightning formation; how a toilet tank works; how ocean waves work; and how a car's braking system works. Two forms of multimedia instruction were investigated: a dynamic-media condition consisting of computer-based animation with accompanying narration and a static-media condition consisting of paper-based illustrations (presented as a series of frames) and accompanying printed text. In both multimedia conditions, relevant principles underlying multimedia design were incorporated (Table 10.2). For example, in the dynamic-media condition, the *temporal contiguity principle* was adhered to by ensuring that the spoken words and associated animations were presented at the same time, and in the static-media condition, the *spatial contiguity principle* was adhered to by presenting printed text near the diagrams that it described. The static-media condition resulted in higher scores in all eight tests (a retention test and a transfer test for each problem topic), thus providing support for the static, as opposed to the dynamic-media hypothesis.

As discussed above, there may be situations where multimedia instruction would benefit from the use of animations, especially when systems or processes are sufficiently complex to preclude mental animation from a series of static snapshots. Also, as was noted, training situations that involve procedural tasks afford advantages if presented, at least to some extent, in dynamic animation form. However, a potential obstacle to learning that is inherent to dynamic animations is that these types of presentations tend to entail the need for *representational holding* (as discussed below), which can induce greater extraneous cognitive load. That is, because dynamic media are transient in nature, dynamic animations with accompanying narration compel holding previously presented information in working memory, which could compromise the amount of cognitive capacity available for essential cognitive processing.

These kinds of concerns pose potentially greater risks to learning by older adults, making it necessary that designers understand the nature of the learning objectives in order to formulate a design strategy. If the instructional topic is too complex to rely on the ability of the learner to animate mentally how a process works from a series of static diagrams, then a purely static-media approach would not be desirable. However, there exist other alternatives that may afford to the older learner a combination of the virtues of both types of multimedia presentations. Specifically, animations could be designed to incorporate positive features of static illustrations through, for example, allowing the learner to control the pace and order of the animations and by cueing important features that might deserve further study following pausing. In addition, by introducing the possibility for interactivity, the learner might be asked to answer questions or make suggestions that could induce more generative cognitive processing.

10.1.3   Multimedia learning and older adults

As implied in the preceding discussion, for older adults a critical consideration with multimedia learning systems is how well the user can process and integrate the various sources of information that are being presented, either simultaneously or successively. Although these instructional environments can, in principle, provide older adults with enhanced opportunities for learning, the discussion on dynamic versus static multimedia presentations highlights how every learning situation presents its own unique set of factors, making it very challenging to determine whether the multimedia system will in fact promote or undermine learning.

For example, an animation of how various parts of a machine work, as opposed to an animation of how to swing a golf club, may be especially more difficult for the older learner. In the machine problem, learning may be taking place by decomposing various actions and parts and trying to retain and integrate this information in WM (i.e., a reliance on representational holding). The golf swing procedure being illustrated by the "human" golfer may, in contrast, evoke a coherent template in memory that could be retained in whole and possibly be used as a basis for emulation.

In general, the uncertainty designers face in developing multimedia instructional environments for older adults stems from a lack of understanding of whether combining sounds, pictures, animation, text, graphics, and interactive controls might be too distracting or overwhelming, or conversely, may actually mitigate some of the cognitive limitations related to attention and memory that occur with age (Chapters 2 and 7). Although it is difficult with the current understanding of multimedia learning for older adults to prescribe the features of all the relevant training-related variables that need to be in place to maximize the prospects for successful learning outcomes, what appears to be certain is that the key consideration is the role of cognitive processing. For older adults, the primary issue that one needs to consider in designing multimedia learning systems concerns limitations in attention that can arise from processing information presented in different modalities and from capacity constraints associated with WM.

As discussed in Chapter 7, it is generally believed that WM has two separate storage systems. One storage system is used to hold and process visual imagery or pictorial information, for instance, where the reset button is located on a picture of a medical device or the direction one must turn a handle following a demonstration of how to activate some device. The other storage system is used to process verbal information, which could be presented visually in the form of text or through the auditory channel in the form of narration. More specifically, the limits in attention that are of most concern in multimedia systems are those associated with the auditory and visual input modalities, and with the pictorial and verbal storage systems in WM.

These limits are especially evident for older adults who may be experiencing declines in the capacities of these information-processing systems, as well as in the speed with which information can be processed (Chapter 7). The existence of these limits provides the basis for many of the guidelines discussed below and presented in Table 10.2. These guidelines have been compiled and adapted from a variety of sources, including Mayer and Moreno (2003), Moreno (2006), Clark and Mayer (2008), and Fisk et al. (2009). For example, such guidelines suggest using both auditory and visual channels to ensure that one of these channels does not become overloaded, using both pictorial and verbal information to ensure that the individual WM storage systems do not become overloaded, presenting information at a reasonable pace and in a way that promotes integration of pictorial and verbal information in WM, and minimizing the confusability of information.

In addition to the visual and auditory channels, multimedia systems could also involve other sensory channels, such as the tactile sense, to account for manipulating various input devices as might be needed to control the pace of an online learning course. To the extent that these devices are unfamiliar to the user, they will exact demands on attention by creating the need for time sharing among multiple information channels, and thus take away attention needed for processing critical learning information. This is why it is so important, as explained in Chapters 4 and 8, to ensure that the older adult learner is sufficiently skilled or practiced on repetitive aspects of the task such as how to control playback of learning material. Otherwise, the older adults' limited processing capacities can be further compromised.

For meaningful learning to take place, the learner must pay attention to the presented material. This implies being able to select relevant auditory and visual information at potentially any point in time and being able to attend to multiple sources or types of information. A further requirement for meaningful learning that may be less obvious is the ability for the learner to integrate and organize the material from these channels into a coherent representation. The learner will also need to maintain and manipulate that information in WM so that it can become connected (i.e., associated) with relevant existing information in long-term memory (LTM). Prior knowledge in LTM is thus critical for guiding the process of both selecting relevant information that is being presented for further processing in WM and for integration of this information (Chapter 7).

As indicated in Chapter 8, this is where supportive information presented prior to practice on a task or past knowledge about the learning material can be critical. Not only can it direct the selective attention process to important and relevant information, but it can facilitate WM processing and transfer of information to the LTM storage system. Once the representations derived from WM integration and organization become schemas or mental models in LTM (Chapter 7) the learning process becomes more efficient, as these schemas and mental models can serve to guide the processing and integration of new incoming auditory and visual information.

Thus, when multimedia instructional systems are to be designed for learners, and particularly for older learners, they need to be carefully constructed so that they are consistent with older learners' information-processing abilities. Age-related declines in sensory and motor abilities also need to be considered. The 10 guidelines for instructional design using multimedia presented in Table 10.2 mostly address information-processing considerations, but consider other factors as well. Although these guidelines can be applied to numerous types of multimedia instructional programs, including many everyday types of learning situations or tutorials that do not involve performing complex tasks, for more involved problems such as learning how to use a software application, emphasis also needs to be given to instructional methods associated with promoting the learning objectives (Chapter 8). Examples of important aspects of the instructional program that may need to be considered and integrated into a multimedia instructional setting might include the design of meaningful holistic practice tasks, the use of scaffolding strategies, the presentation of supportive and just-in-time information, and scheduling practice sessions to minimize fatigue.

For older adults, multimedia might be especially useful at the beginning of an instructional presentation when there may be some anxiety regarding confronting new information or concepts (Chapter 6). Having a narrator's personalized style (see Table 10.2) serve as a calming influence may help to allay such fears. For example, the presentation of a computer-assisted drawing program can begin by slowly displaying a progression of different types of output that the learner will be able to learn how to produce, with the narrator's voice informing the learner that although these drawings may seem complicated, "We will make the process of learning how to produce them relatively easy to grasp so that in no time you will be doing it."

The narrator's personalized style can also be used to help direct the learner's attention to visually presented information through appropriate emphasis and directives provided in the narration. Specifically, emphasis in speech can be used advantageously to direct the user's attention when providing supportive information needed for subsequent demonstrations of worked examples and the performance of practice tasks. For example, for online instruction in PowerPoint the narrator can direct the learner's attention to the left-hand side of the screen, where the outline view may be demonstrated, and then to the right-hand side of the screen, where the slide view is presented. The narrator can then emphasize how the typing of text in the outline view, which is being simulated, translates into the output in the slide view, and then, using a new slide, how the input of text directly into the slide does not have the same effect on the outline view.

Unimodal presentations can provide this same information, with arrows, highlighting, and other figural artifacts being integrated into the presentation to emphasize the various considerations associated with the two modes of creating slides. However, the benefits of having a narrator guide one's attention to this relatively simple yet important concept should not be understated. Aside from the obvious advantage that the user does not have to split attention between the explanatory text and the slides that are unfolding, the appearance of the integrated or overlaid pictorial and textual information can be timed to coincide with the narrator's prompts, thus allowing for more focused processing of information. As noted in Table 10.2, it is important not to overload the two modalities when integrating narration and visually presented information, including text. Thus the appropriate pausing of narration, to give the learner the opportunity for processing what might be an abundant amount of screen information, as well as the appropriate times to insert narration and the amount of narration to use at any given time may need to be carefully thought out.

Likewise, many other aspects of instruction on this computer software application can benefit from multimodal presentation of material. For example, the process of systematically exploring various functionalities provided in the PowerPoint menus can benefit from the use of narration and animation. The animation can be in the form of simulations of menu selections and corresponding displays of resulting pictures, text, and changes in the design of the presentation. Narration can serve to reinforce the mental model or concepts underlying the menu item in question and focus the learner's attention on the relevant aspects of the functionality, with the learner able to pause or repeat the instructional segment. As implied above, the success of the multimedia platform will depend on the nature of the to-be-learned material, and for complex tasks will require the integration of learning principles from Chapter 8.

10.1.4   *Interactivity in multimedia instruction: Revisiting the issue of active versus passive learning*

In Chapter 8, the matter of whether being actively involved in the instructional program as compared to assuming a more passive role was considered in terms of the effectiveness of learning outcomes. The implications from that chapter were that active involvement by the learner was not necessarily advantageous; under certain types of learning conditions it could actually increase extraneous cognitive load due to additional information that may need to be processed or attention becoming divided in ways that can disrupt information processing.

This issue, however, is far from straightforward. Active engagement as opposed to passive learning has often been recommended, particularly in human factors circles, as a means for inducing deeper learning and improved retention. Generally, the literature has shown positive, negative, and mixed results with regard to the effects of interactivity on the quality and effectiveness of the learning environment.

The divergent results regarding the effects of interactivity could be due to different views in the various studies on what interactivity means or entails. For example, in multimedia learning, does interactivity relate to the ability for the learner to control the pacing or sequence of instruction or the start, stop, and speed of video or animation segments? Does it also relate to whether the learners are given the opportunity to control their responses and system feedback, or organize the instructional content of the learning materials? There are many other different forms of interactivity, especially in multimedia environments, that could be considered as well, and these distinctions could very well account for the different findings regarding the potential learning benefits associated with engaging the learner through interactivity.

Domagk, Schwartz, and Plass (2010) have examined this issue in detail and arrived at the following definition of interactivity: "Interactivity in the context of computer-based multimedia learning is reciprocal activity between a learner and a multimedia learning system, in which the [re] action of the learner is dependent upon the [re]action of the system and vice versa" (p. 1025). Thus, the emphasis is on the "dynamic relationship between the learner and the learning system." Traditionally, approaches to defining interactivity have focused on either: a technological perspective, which emphasizes the affordances provided by the system to engage the learner in various ways (e.g., use of input devices such as a touchscreen, mouse, or keyboard; immersive simulations; or more generally, in the many ways in which the learner can manipulate the learning material); or a psychological and learner-centered perspective, which may incorporate aspects of affordances of technological systems (e.g., manipulating control over a presentation) but that focus on the cognitive processes in the learner that these technological manipulations can affect.

However, Domagk et al. (2010) argued that despite calling attention to the learner's cognitive processes, the types of interactivity that emphasize to a greater extent the psychological perspective to interactivity are still not sufficiently separable from the technological perspective to interactivity. Instead, they proposed a model of interactivity between the learning system and learner that does not emphasize the affordances of the learning system alone or the cognitive activities of the learner, but rather an integration of both. This integrated model of multimedia

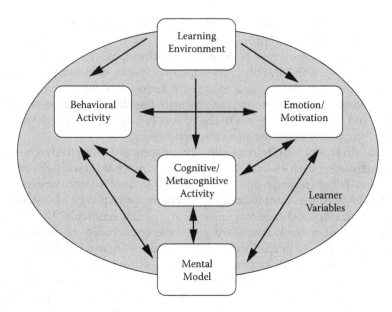

Figure 10.1 The integrated model of multimedia interactivity (INTERACT). (From Domagk, Schwartz, and Plass, 2010. *Computers in Human Behavior, 26*: 1024– 1033. With permission.)

interactivity (INTERACT) consists of six principal components as depicted in Figure 10.1; each is discussed in detail by these authors.

Here we emphasize the relevance of some of these components for older learners in multimedia learning settings. Referring to Figure 10.1, the learning environment, which includes the instructional design as well as the affordances of the learning system, emphasizes the feedback loops and a unidirectional loop that connect the learning system, behavioral processes, and cognitive processes. It also distinguishes between what the system affords (e.g., the ability to move back and forth), which is a feature of the learning system, and what the learner actually does (e.g., touching a screen, entering text, or clicking on a button), which represents the behavioral activity component of the model. The learner's actions can induce changes in the system that could, in turn, lead to changes in the learner. This, in fact, represents one of the challenges in designing behavioral affordances for older learners in multimedia environments: which types of behavioral activities are most likely to have a positive impact on the learner by virtue of the changes that these activities induce in the system?

For example, there is some evidence that allowing for exploration of a virtual environment by using a mouse to freely control one's path could lead to better spatial learning as compared to a passive viewer exposed to the same spatial information (Péruch and Wilson, 2004). With older adults, the cost effectiveness associated with behavioral affordances needs to be evaluated. How much effort would it take for the older learner to learn how to use, or recall how to use, the particular interactive feature and how might this feature disrupt processes of attention? Also, what benefits to understanding and deeper learning can the outcomes of these behavioral interactions possibly provide?

The cognitive/metacognitive component of the model refers to how interactive features can promote cognitive processes such as remembering, understanding, applying, analyzing, evaluating, and creating. These processes are linked to the different dimensions of knowledge, notably factual (i.e., declarative), conceptual, procedural, and *metacognitive*, where the latter refers to knowledge about what we know, that is, knowledge about cognition and activities that can be used to regulate one's cognitive activities and thereby control learning (Chapter 7). The metacognitive strategies usually considered for this purpose are planning, monitoring, and evaluation.

Metacognitive activities can be crucial to positive learning outcomes. For example, by using an effective monitoring strategy, the learner can identify points in time when cognitive activities such as understanding, evaluating, or even remembering are faltering, and accordingly invoke help or guidance from the system. An effective metacognitive planning strategy may allow the learner to determine how much effort to allocate to certain cognitive activities, such as concept formation, in terms of how critical that information may be to overall learning. Metacognitive activities can be very resource efficient in the sense that these higher-order cognitive strategies could reduce the cognitive load associated with those cognitive activities that are more directly linked to processing information and learning. For older adults, the key is to have interactive features available to the learner that induce effective metacognitive strategies, for example, that might prompt the learner for review or rehearsal, for a pause (to allow time for reflection), or other means of inducing higher-level evaluation of the ongoing learning effort. This issue is discussed in Section 10.2.5.

The INTERACT model also acknowledges the potentially important influence of interactive features on the learner's emotions and motivation. Specifically, emotional states that can arise due to the nature of a particular learning situation could impair the learner's motivation and result in negative learning outcomes. The links among the learning environment, emotion, motivation, and cognition are complex but could potentially play an important role in the design of interactivity in multimedia learning environments, especially for older adults who may be more prone to confusion from input devices or problems that may surface from a lack of smooth operation in the interactive features. These factors, along with other model components (Figure 10.1), can affect the learner's mental model, which is continuously being developed and possibly revised during the learner's integration and organization of new information.

This model includes, in the backdrop of its other model components, the very important consideration of learner variables (Figure 10.1). These comprise cognitive and metacognitive characteristics such as degree of prior knowledge, and affective (nontransitory) traits such as self-efficacy and trait anxiety. These variables can affect each of the components in the INTERACT model. With older learners, normative age-related declines in cognitive abilities need to be considered in any multimedia interactive arrangement. In principle, the interactivity has both the ability to induce extraneous cognitive load and compensate for cognitive ability declines through the promotion of effective, resource-efficient metacognitive strategies. Likewise, prior knowledge that the older learner brings to the learning situation can contribute to more effective learning if the learner can draw from a wealth of lifelong knowledge (Chapter 4) as a basis for drawing analogies or metaphors. For older learners, the lack of critical technological knowledge or relevant concepts may cause gaps in the learning process that may be too difficult for them to bridge.

Finally, it should be noted that sometimes interactivity may have counterintuitive effects. This may explain why the "active" learning condition in some studies resulted in poorer learning outcomes than a "passive" condition. However, such results could, in fact, be consistent with the INTERACT model. For example, an "interactive" condition may focus on providing a particular kind of guidance that may, in effect, be less conducive to using deeper cognitive abilities, thereby explaining the lack of effect of interactivity.

10.1.5 Some empirical studies related to multimedia instruction and older adults

To date, there have been relatively few studies that have examined the potential benefits of multimedia instruction for older adults. One study by Van Gerven et al. (2003) investigated the benefits of multimedia presentation of worked examples for older adults (for more on worked examples, see Chapter 8 and Section 10.2.10). In using worked examples for instruction on problem solving, problems are accompanied by the subsequent solution steps that lead to the goal state. In contrast, approaches that require the learner to apply means–ends analysis require the learner to process the problem backward from a set goal. By focusing the learner's attention on problem states as opposed to operations needed to derive solutions, worked examples should impose less cognitive load on the learner, especially during the early stages of learning when the learner has few if any cognitive schemas available for solving the problem (Chapters 4 and 7). This savings in processing capacity early in learning can then be used to form rudimentary cognitive schemas that can, with further learning, be transformed to more useful schemas that are often needed during the course of the learning process.

Consistent with guidelines in Table 10.2, it would seem that the use of multiple modalities, as opposed to a unimodal modality, could enhance the benefits of worked examples for older learners by further reducing cognitive load through the distribution of information over the different limited-capacity WM stores. Van Gervin et al. (2003) examined this use of multimedia-based worked examples by presenting pictorial information in animated form and the accompanying text in narrated form, while ensuring temporal contiguity of the visual and auditory information (see Table 10.2). In addition, these investigators hypothesized that animation would have benefits over nonanimated sequential (i.e., static) presentations of the worked examples (that had explanatory text integrated with the images) because it would avoid "repetition of invariable pictorial elements." Thus, consistent with cognitive load theory, it should reduce extraneous load to which, as noted in Chapter 7, older adults are more susceptible.

These hypotheses were tested using a computer-based problem-solving task performed by a sample of younger and older participants under three conditions: a conventional problems (CP) condition (i.e., a non-worked examples condition), a unimodal worked examples (UWE) condition in which the training examples were presented in the form of a static sequence of pictures, and a multimedia-based worked examples condition (MWE), where the training problems were presented in an animated form and accompanied by a narrated (as opposed to a printed text) explanation. In addition to performance results that were based on a series of test problems, data were also collected on subjective cognitive load (SCL) as perceived by the participants during the training exercises.

The SCL findings indicated that both age groups experienced lower levels of cognitive load in the MWE condition than in the CP condition; the MWE condition did not reduce the SCL relative to the UWE condition; and that the older participants demonstrated a greater decrease in SCL between the CP and the MWE condition as compared to the younger participants. The performance data revealed that the younger participants performed significantly better than the older participants. In comparing the CP to the MWE condition, there was a trend toward better performance in the MWE condition and disproportionately so for the older participants. No training format or training format by age interactions were found in the comparisons between the UWE and MWE conditions or between the CP and UWE conditions. Thus, overall there seems to be some evidence for benefits of multimedia, especially in comparison to a conventional training approach to problem solving. However, these results certainly do not provide any compelling argument for the benefit of multimedia presentations of worked examples, especially for older adults.

An area related to multimedia is the idea of *multimodal interaction*, where multiple sensory input and output channels can be used when interacting with a technology. The multimodal design strategy offers users the choice of which modality they want to use, whether they want to combine modalities, or whether they want to alternate between modalities. Although the emphasis in multimodal interaction is on performing a task, its relationship to multimedia instruction lies in its ability to have a variety of modality options available to the user in training or instruction scenarios.

Obviously, there are many factors that might influence a user to choose one modality over another when performing a task, including the nature of the particular task activity, the environmental constraints, and the familiarity, comfort, or abilities of the user with regard to particular modalities. In work or other task settings, the choices inherent to multimodal interaction provide for a greater degree of inclusiveness: for example, people with motor impairments can input information through speech rather than through scrolling and selecting items from menus, or have the opportunity to switch to another modality following the recognition of an error or difficulty associated with the previous modality. The ability to shift to different modalities has also led to the suggestion that multimodality could reduce cognitive load through the corresponding distribution of mental load to different modality channels (Oviatt, 2003). However, as implied from the discussion in Chapter 7 on executive control processes needed for integrating spatial and verbal information in WM, multimodality could also increase cognitive load due to the extra mental resources needed for evaluating a current modality's effectiveness and the need to switch to or coordinate with other modalities.

The question of whether older people may benefit more than younger people when given the option to use single modalities and multimodality was investigated in a study by Naumann, Wechsung, and Hurtienne (2010). Participants used a specially configured smart phone with touch, speech, and motion input to solve a series of tasks (e.g., accessing voice messages, viewing e-mail, and redirecting calls to a default number). They performed the tasks in three single input modality conditions consisting of either touch, speech, or motion control, followed by a fourth multimodality condition in which the participants could switch or combine modalities in any way that they desired. In addition to performance data, subjective data related to user satisfaction, including perceived cognitive load and perceived effort for learning, was collected.

As expected, the performance data were more favorable for the younger participants; they demonstrated better performance and shorter task durations than their older counterparts. Generally, the multimodal and the touch condition were rated more favorably than the speech and motion conditions, and speech was rated as better than motion. For a questionnaire subscale referred to as *hedonic quality-stimulation* (perceived fulfillment of the need for developing one's knowledge and skills), the multimodal condition was rated better than the touch condition. Also, the older participants rated the motion control input modality worse than the younger participants. Given that the older people were also less successful when using motion control than their younger counterparts, at least for this task it would appear that this modality was the least appropriate for them, which is consistent with evidence related to the deterioration of movement skills with increased age.

With regard to the choice for a modality in the multimodal condition an age group difference was not observed, with the touch input modality being the one most preferred by both age groups. However, the older participants showed a less flexible interaction strategy than the younger participants; after a failed attempt using an input modality, the older people were less likely to switch the modality as compared to the younger people. Overall, these results provide some evidence that older users may not prefer to switch modalities and would rather keep using a modality they were comfortable with or felt confident using. This strategy may reflect a form of metacognition, whereby older users are aware of the increased mental effort involved in making choices between different modalities, and thus may lack confidence in making use of the flexibility of multimodal interaction. From the standpoint of multimedia training, a lesson that may be learned from this research is the need to ascertain the preferred input modalities of older learners and to ensure that they are facile on these input modalities if they are needed as part of the interaction with the multimedia system. This information can then be used to dynamically adapt the training system to the individual's modality preferences, assuming the multimedia learning platform is sufficiently flexible.

The use of multimedia training as a means for helping older users negotiate the U.S. government's Medicare.gov website was recently examined by Czaja et al. (2012). In this study, the use of multimedia training was not applied in the more conventional way, that is, as a means for learning about a particular domain or tailored to solving a particular type of problem. Rather, its focus was on presenting a broad training tutorial intended to help orient the user to the website's homepage and facilitate navigation throughout the website prior to engaging in any of the array of problem-solving activities that this website could support.

Toward this end, the tutorial organized the homepage into four general *windows into the website* (Figure 10.2a), which were then further elaborated on. The training tutorial also provided general information that included *tips* (e.g., make sure to scroll to the bottom of the page), *concepts* (e.g., drop-down menus, links, buttons, and check boxes), *tactics* (e.g., in performing searches using search boxes), and *metaphors* (that were used to partition the homepage into more easily discernable and understandable categories and thus support the generation of a mental model of the homepage).

(a)

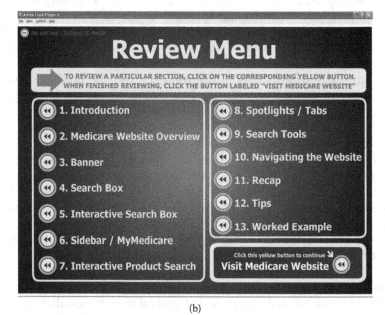

(b)

Figure 10.2 (a) The Medicare.gov home page and subparts. (b) Review menu for multimodal segments.

In this study, older adults were randomly assigned to one of three conditions. The participants in a multimodal (multimedia) condition viewed the training module in a format that made use of two sensory modalities, visual and auditory, within the context of an Adobe Flash presentation that allowed use of its animation features. The training was divided into segments or modules, and each module could be reviewed as many times as desired. Following the completion of the training tutorial, participants were presented with a menu containing links to all of the modules and could choose to review any of the module segments as many times as needed (Figure 10.2b). The animated details that were enabled through the multimodal presentation were intended to provide a more salient illustration of selected ideas and concepts. The narrated audio component that was used to inform the learner about the material was complemented by onscreen text which highlighted important features in the narration. In addition, the multimodal condition made use of interactive examples: participants had to use various items or objects that they might ordinarily encounter or interact with during actual problem solving with this website (e.g., typing words into search boxes, activating dropdown menus, clicking onscreen buttons).

In the unimodal condition, this information was presented in purely visual form, within the context of a Microsoft PowerPoint presentation. In addition to graphics, each PowerPoint slide also contained relevant text and callouts which described or amplified the slide's topic. The unimodal PowerPoint presentation was adapted from the multimodal Adobe Flash presentation and contained very similar, if not completely identical, information. In a third condition referred to as "cold start," participants did not receive training. Instead, they were allowed to browse the Internet for a period of time (30 minutes) that was approximately equal to the training time associated with the other two conditions. Following training or browsing, participants were required to answer three relatively complex multistep problems using search tools available on the Medicare.gov website. Three measures of performance were computed for each task: a problem-solving accuracy score; an efficiency measure, which was the accuracy score for the task divided by the time taken to perform the task; and a measure of navigation, which reflected how far into the "problem space" (each task was divided into stages) the participant was able to reach in the task. Average measures of accuracy and efficiency were computed based on performance across the three tasks.

The results demonstrated a trend in the data indicating that those in the multimedia condition had higher accuracy in their performance than those in the other two conditions. Also, a significantly greater number of participants in the multimedia condition were able to traverse farther into the problem space for one of the tasks, with a trend for this effect observed for one of the other tasks. The analysis also revealed that even after controlling for the amount of prior Internet experience participants had, both fluid and crystallized cognitive abilities (Chapters 2 and 4) were very strong predictors of accuracy and efficiency and that training (reflected by those who received either unimodal or multimedia training) still had a significant effect even after cognitive abilities were taken into account. Thus, this study did provide some evidence for the effectiveness of training, and particularly multimedia training, in negotiating the complexity of virtual spaces consisting of interlinked web pages.

10.2   e-Learning

10.2.1   e-Learning environment

e-Learning represents a computer-based method of instruction or training that is typically delivered through the Internet and corporate intranets, and to a much lesser extent through CD-ROMs. Because it can facilitate active and independent online learning, where users are responsible for their own learning process, it has become a popular alternative to face-to-face training. e-Learning embraces multimedia in the sense that it uses various media elements such as narration, text, pictures, and videos to deliver the instructional content.

Within e-learning, a distinction can be made between *asynchronous* e-learning and *synchronous* e-learning. Asynchronous e-learning is designed for self-paced individual self-study. In contrast, synchronous e-learning is typically instructor-led and designed for group learning in the form of virtual classrooms, which are being increasingly used to build knowledge and skills linked to improvements in job performance. Synchronous e-learning possesses some common features with asynchronous e-learning. In particular, it relies on computer screens to communicate content and apply instructional methods and also benefits from interactions with learners to maintain attention and promote learning. However, synchronous e-learning, like conventional instructor-led face-to-face classrooms, also enables a high degree of social presence through, for example, screen chat boxes that can allow each participant to type messages, or screen audio control boxes that can be used by the instructors or participants when they want to speak. Headsets equipped with microphones enable the instructor and participants to speak to and hear one another. Also, like conventional instructor-led classrooms, the presentation rates of the learning materials are not controlled by the learners. Either the synchronous or asynchronous format can support asynchronous collaboration with other people, for example, through discussion boards, blogs, wikis, and e-mail.

Because the possibilities when interacting with an instructional or training system online are vast, e-learning is perhaps better viewed as a learning environment. This environment may often need to be navigated through to obtain instructional material, not unlike how we navigate online in pursuit of informational resources. When e-learning's synchronous features are available, this environment may be needed for communicating with other individuals who may consist of similar learners, trainers, or other types of people.

10.2.2 Instructional methods unique to e-learning

As emphasized in Chapter 8, learning is largely governed by the instructional method used for the particular learning task. This is even more likely to be the case with older learners. However, what needs to be recognized is that certain media offer unique opportunities for delivering instructional methods, and in this regard media could play an important role in the potential for effective learning. Clark and Mayer (2008) have noted four instructional methods that are relatively unique to e-learning and which can potentially have a favorable impact on learning, although the impact may be larger for asynchronous as opposed to synchronous e-learning.

One method concerns the management of practice and feedback, two features critical to older learners. For example, following an animated narrated demonstration of how to insert an object, such as a photo, into a PowerPoint slide, the course can direct the learner to perform the same or a similar operation. With asynchronous e-learning, the learner's actions can be evaluated, leading to hints or feedback in support of detecting what was done incorrectly and how to correct any problems with execution of the procedure. Because synchronous e-learning involves actions from a number of learners, the nature of the evaluation will be based on a review of actions from a group of students, and the corresponding feedback will resemble more closely what is provided in a traditional classroom.

Closely related to these practice and feedback features is the method of customizing or tailoring instruction. The discussion concerning situated cognition in Chapter 8 hinted at the need to adapt instruction to the way the older learner perceives the task or problem context. This could include the kinds of metaphors the older learner may be more comfortable seeing applied to the situation, or even the sequence in which material should be presented. The ability to make ongoing dynamic adjustments within the path of the instructional program based on the responses of learners is a hallmark of any good asynchronous e-learning program, and is often referred to as *adaptive instruction*. Examples of adaptive instruction include evaluating the nature of the error(s) made by the learner and the level of complexity of the problem as a basis for the kinds of practice problems that will be subsequently presented to the learner. For example, the instructional program may determine the need to present examples that are slightly less complex but that still contain the opportunity for making the same kind of errors. In addition to practice problems, e-learning presents the opportunity for dynamic adjustment to be made to worked examples. In Chapter 8, Section 8.4, "Worked Examples and Problem-Solving Exercises," the use of worked examples as a method of instruction was discussed within the broader topic related to the sequencing of instruction. Later on in this chapter, worked examples are discussed within the context of an e-learning instructional platform.

A third instructional method that can be easily supported by e-learning is the ability for social collaboration, for example, whereby learners can collaborate at independent times through discussion boards or other collaborative tools. Although the conditions under which learning outcomes can be improved through interaction or working together with others are still not clear, there are some limited guidelines beginning to emerge regarding ways to use the Internet to foster collaboration that can facilitate learning. This may be an interesting area of research with regard to older learners, especially those who may feel more isolated than younger learners. Such learners may be able to harness the social rewards associated with collaborative learning into a positive learning asset.

A fourth instructional method relatively unique to e-learning involves the use of simulations and games. The motivational appeal of many online games has triggered an interest in constructing learning games based on software simulations. This type of learning environment is conducive to situated learning, as it can capture the dynamic complexity of many actual situations in which a person must evaluate various factors for the purpose of making a decision. It is also consistent with principles underlying the 4C/ID model (Chapter 8) in the sense that holistic tasks at various levels of complexity can be systematically presented to the learner.

The tasks in simulated game environments, however, are more likely to be directed toward relatively unique types of learning experiences. For example, jobs in the banking and insurance industries involving providing financial advice to clients, recommending funding for commercial loan applicants, or explaining different insurance options to prospective customers can be simulated through various objects that constitute realistic job situations. Initially, the job problems can be made very simple in order to help establish the older learner's confidence and to minimize the number of factors to which the learner may need to direct attention. Key objects, such as a computer database that may need to be accessed for processing a customer's qualifications for a loan, could be highlighted early on but could then be made less apparent as the learner develops expertise. Adaptive instruction as applied to simulated games could greatly benefit older learners. Depending on how these learners discern the situation and make decisions, appropriate feedback and practice simulation exercises at lower levels of complexity may become triggered, with emphasis on the facts and concepts that were problematic for the learner. Narrated

demonstrations could also be incorporated during the initial job problems, consistent with the 4C/ID model, to help focus the older learner's
attention on key aspects of the problem. Likewise consistent with this
instructional model, just-in-time practice on specific skills such as how to
search a database can be given as well, all within a scaffolding perspective
(Chapter 8) that enables the learner to gradually become more independent in performing the tasks.

10.2.3   Some concerns with e-learning

There are two potential dangers associated with e-learning systems.
First, there is the danger that instructional designers may place too much
emphasis on the technology and less emphasis on the task-specific job
or task domains in which they are trying to provide expertise. A rigor-
ous task analysis is thus essential for ensuring that the knowledge and
skills that underlie the needed expertise are indeed being captured by the
e-learning instructional platform. Second, as implied in the discussion of
multimedia, the numerous software features afforded by e-learning make
it easy to lose sight of human limitations, especially the cognitive capabilities of older learners, and thus possibly overload the learner's information-processing capacity.

Conversely, there is the concern that instructional designers may opt
to convert traditional learning environments, such as books or face-to-face
classroom presentations, into their e-learning analogues (e.g., the pages in
the book are now turned using software controls), without taking advantage of the unique features of e-learning. Instructional designers need
to search for a balance between these two extreme approaches to implementing e-learning platforms.

These concerns notwithstanding, the popularity of e-learning as an
instructional platform in the corporate world continues to grow. This can
be attributed to a host of attractive features, including some of the unique
aspects of e-learning in providing instruction that were noted above, such
as its ability to be presented at the learner's own pace, enable easy access
to additional relevant resources, and be tailored toward the learner's level
of skill and knowledge and customized to the learning styles and preferences of different types of users. Its practical value should also not be
underestimated: e-learning can eliminate many training costs related to
travel and space, and can, in principle, be made available any time and
any place.

10.2.4   Challenges in designing e-learning systems

The design of e-learning systems is governed by considerations associated
with three broad and interrelated areas:

1. Human–computer interaction (HCI), and particularly, the broader aspects of HCI that concern the computer or graphical user interface (GUI) which serves as the intermediary between the learning platform and the user.
2. Adaptive control, which concerns the specialty area of HCI that addresses the issue of how the instructional system adapts to the user's prior knowledge, skills, or needs, and also includes the degree to which decisions regarding instructional support are shared between the learner and the system.
3. The manner in which multimedia content is presented within this interface.

Obviously, this is a tall order for designers catering to the general population of users interested in e-learning. It would require, assuming designers do not want to count on their intuition alone, invoking scores of principles of human–computer interaction that are capable of affecting learning in this environment, and integrating these with a number of principles of multimedia design for instruction. Given the pressures that these designers are likely to face in terms of the resources needed to ensure that the designs are adequate (e.g., to ensure participatory design and appropriate testing and evaluation), in addition to the frequent need for product delivery in the face of tight schedules, many are apt to rely on common sense or borrow designs that were acceptable for other learning objectives. The task of designing such e-learning environments for older learners is especially daunting, as the number of design considerations that would have to be taken into account is extensive. These powerful learning environments have the potential to confound older users in numerous ways, especially users who are not facile with graphical user interfaces.

Design issues related to HCI that are of a broader nature and may be relevant to older users constitute a topic which is covered elsewhere (e.g., Fisk et al., 2009). The implications of multimedia design for older users were covered earlier in this chapter. In the next section we turn to the issue of adaptive control, which is particularly relevant to the design of asynchronous e-learning systems. Synchronous e-learning systems, by virtue of being instructor-driven, leave little room for learner control and thus negate opportunities for older learners to manage potential cognitive overloading situations. Asynchronous systems, however, although capable of providing the flexibility for such offload, can present other challenges for older adults who need to confront the technological features of the system and its interface in "isolation." Some of the issues related to asynchronous e-learning systems and older adults are covered later.

10.2.5   *Adaptive and shared control in e-learning systems*

People who are familiar with the way one navigates on the Internet and with the control features that promote such browsing behaviors will be more likely to feel there is something amiss in an asynchronous e-learning system that does not provide similar control freedom. Three fundamental categories of control options that e-learning systems can easily provide and should offer to the user are: (1) control of the sequence in which the learning content is delivered, (2) control of the pace at which the learner wishes to examine the learning materials, and (3) control of access and management of various learning support elements such as worked examples and practice exercises, glossaries, help or advice systems, reference materials, and links to websites.

By providing learners with options for selecting the topics they may want to study next, the pace at which they want to progress through the learning materials, or the support system such as a coaching system or a link to an additional resource on the web, there is an implicit assumption that at least some learners will benefit from the choices that these kinds of control options afford. This, in turn, presumes that learners are good at self-assessments regarding what they know and what they need to do to close any gaps in their knowledge about a topic.

Unfortunately, there is evidence that these subjective assessments are more illusory than factual (Glenberg, Wilkinson, and Epstein, 1992). People with high metacognitive skills, however, should be expected to be more effective at managing their learning activities, presumably because they can more accurately judge what knowledge they are lacking and how they should allocate their information-processing resources to close the gaps in their knowledge. This would include which control options they should choose in an e-learning system to help them meet the knowledge and skill acquisition objectives of the instructional program.

Do older learners have better metacognitive, and thus, learning management skills? This is a difficult question to answer, however, some insight into this issue can be gathered indirectly. For example, based on managers' assessments of work-related attributes of older as compared to younger workers, older workers were perceived as better than younger workers in terms of time management ability, working independently, and maturity (Sharit et al., 2009). Also, older people generally are more self-aware of limitations, given normal age-related declines that occur in cognitive abilities (Chapters 4 and 7). Such considerations suggest that older learners may be more aware of their limitations in knowledge and skills, and possess the maturity and discipline that would make them more likely to seek e-learning interface control options that can reduce the gaps they perceive themselves as having with regard to the learning topic. Building adequate learning control into the instructional system is thus essential for older learners who may be less illusory in their self-assessments of their knowledge as compared to younger learners, and more likely to effectively utilize control features.

With respect to the design of learner control features, Clark and Mayer (2008) recommended four guidelines. The first guideline addresses the issue of trading off learner control with program control: should control over the learning materials be handed over to the learner or should it be primarily dictated by the e-learning system? As implied above, people with good metacognitive skills, and we assume that older adults generally have such skills, would benefit from learner control. Learner control is also more conducive to providing a more interesting, and thus perhaps more motivating learning environment. Other factors that favor learner control are when learners have prior knowledge of the instructional content, when the learning topic is of low complexity, or when the training represents a more advanced component or module of a training program.

One concern in applying this guideline to older adults is that this population of learners, although they may possess good metacognitive skills, may also have less familiarity with the topic, particularly if the topic involves knowledge regarding rapidly evolving domains such as technology. This concern relates to a second guideline concerning learner control: people in the early stages of learning or those who have low prior knowledge of the subject matter may tend to use less effective learning strategies (e.g., choosing an inappropriate sequence of worked examples) when given the degrees of freedom that learner control affords. These learners may benefit more from decisions made by the program.

With older learners it is probably best to trust their metacognitive skills to make good decisions with regard to the various control features the e-learning system offers. However, to ensure that they do not disregard worked examples or practice exercises essential for acquisition of fundamental knowledge and skills, these features can be placed under program control or at least made difficult to be bypassed by the learner. Also, as emphasized in Section 10.2.7, it is essential that additional time be allocated to older learners to enable them to become familiar with the learner control features. This strategy is important as lack of familiarity or uncertainty regarding these features can disrupt learning by forcing essential cognitive load to be directed toward comprehension of auxiliary control functions. By training older adults on the use of these control elements to the point of automaticity (Chapter 7), metacognitive skills can be more easily made manifest.

A third guideline in the design of control features in e-learning systems, and one that is critical to older learners, is to provide as much pacing control as possible. By virtue of the provision of pacing control, asynchronous learning environments can be much less cognitively overloading than their corresponding synchronous learning environments. Pacing control refers to a variety of features that control the pace at which learning proceeds. These can include use of the "forward" and "back" buttons, "replay" buttons for video presentations, and "quit" and "continue" options. It is essential that older learners be aware of where all functions that can affect the pace of learning are situated, as these functions are likely to be the older learner's best friend, their closest and most accessible allies in their attempts to shield themselves from cognitive overload.

Also affecting the pace of learning, but mostly by aiding the learner's ability to perceive, organize, and comprehend the learning experience, are an array of navigational features. Many of the guidelines dictating the design of these features are similar to those that exist for the design of websites, where users seeking information may confront challenges that are similar to those faced by learners interacting with e-learning systems. These features include topic headers, topic menus, course or site maps (which graphically represent the topics or structure of a training module or lesson), and links to other resources. Headers and labels should be salient, and meet guidelines related to perception discussed in Chapter 7. Links should be used with caution, especially with older adults, as leaving the primary instructional area to gather additional instructional material creates both a temporal and spatial separation that violates temporal and spatial contiguity principles in multimedia design (Table 10.2), and thus potentially increases extraneous cognitive load on the learner by placing the burden on the learner to integrate this information over time and space.

10.2.5.1 Adaptive control

A fourth set of guidelines in the design of control features in e-learning systems relates to the very important concept of *adaptive control*. This concept does not just apply to learning systems but to a broad class of systems with which humans interact, and thus represents an important topic within the area of HCI. Broadly, in a human–computer interactive system adaptive control considers the degree to which the automation or programmed components of the system recognize the nature of the human's performance within the context of the system's state and performance goals. Based on this assessment, the automation dynamically adjusts control of the system, for example, by taking over certain decision-making responsibilities previously accorded to the human controller, or by advising the human about which actions should be considered. In the field of human factors, this area has been referred to as *allocation of functions*. In principle, decisions regarding the allocation of functions can span from a static allocation policy to various levels of dynamic allocation of decision making and control.

Within the context of e-learning systems, it is the learner's responses to the instructional program that trigger the adaptive system. The adaptive system would then consider adjustments that should be made to the presentation of learning materials or other forms of support directed at the learner.

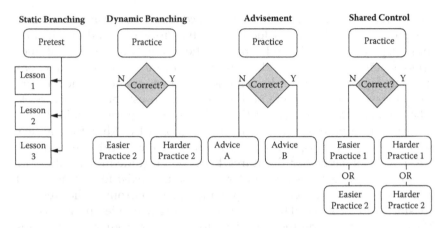

Figure 10.3 Four forms of adaptive control. (From Clark and Mayer, 2008. *E-Learning and the Science of Instruction: Proven Guidelines for Consumers and Designers of Multimedia Learning*, 2nd ed. San Francisco: John Wiley & Sons. With permission.)

For example, if the learner appears to be doing well on practice tasks, the program could use this information to present the learner with a more challenging problem than would otherwise have been presented. Conversely, the detection of poor performance may signal the need for the program to retreat back to simpler problems or ones that cover fundamental ideas that seem to be at the core of the problems that the learner is experiencing.

As in the broader area of allocation of functions, adaptive control in e-learning systems can range from *static adaptive control*, which in reality is not truly adaptive, to various forms of dynamic control strategies. Four generalized forms of adaptive control in e-learning systems are depicted in Figure 10.3. As noted, on one extreme is static adaptive control. This strategy relies on some prior measure of learner knowledge or skill as a basis for branching to different types of instruction, but the presentation of the learning as determined by this prior assessment is static or predetermined. This strategy can actually benefit older learners in several ways. For example, if the learning system contains technological features that some prior assessment may suggest would be troubling for the older user, or terminology that may be unfamiliar, the program may branch to an interface or lesson that is more conducive to the user's level of technological expertise or knowledge.

The other three general categories depicted in Figure 10.3 reflect forms of adaptation that are more dynamic in nature. In *dynamic adaptive control*, which is the second form of adaptive control shown in Figure 10.3 (referred to as dynamic branching), the number and complexity of practice tasks may be adjusted based on the learner's performance on previous practice tasks. This is in contrast to a *fixed program control* strategy whereby the learner is presented with a predetermined set and sequence of practice tasks. Although there is no strong evidence that such adaptive control can lead to higher performance levels, it may result in substantially reduced time and effort needed to achieve these same performance levels (Salden et al., 2004). This potential advantage in efficiency, which can translate into reduced investment by older learners in their cognitive resources, makes dynamic adaptive control a strategy that should be strongly considered for older learners.

The third form of adaptive control is a type of dynamic adaptive strategy referred to as *advisement* in Figure 10.3. This type of adaptive guidance occurs when the program determines the need for providing advice to the learner. In terms of allocation of functions, this approach represents a lower level of program or automation control and allows the learner to make the decision regarding whether to heed the program's advice. For example, based on the learner's performance on practice problems the program may suggest that the learner review a particular module, perhaps because the learner's performance implied a lack of understanding of a particular concept.

Adaptive guidance does not replace learner control, but rather superimposes on it an advising component. This is likely to lead to longer lesson times due to engaging in more problems or spending more time studying certain topics, but has also been found to result in better performance, particularly on far-transfer tasks, than those in the standard e-learning learner-controlled program. Adaptive guidance programs may also be more suitable for older learners as they incorporate a friendly, more sociable component into the interactive learning environment, which older learners may find less intimidating. The real challenge, of course, is working out the logic underlying the advice or recommendations that the program seeks to present dynamically to the learner. If there is a desire to train older workers on tasks deemed important, the investment in developing and validating the logic underlying the adaptive guidance program may be cost effective.

The fourth type of adaptive control depicted in Figure 10.3 is *shared control*. This strategy represents a hybrid of dynamic adaptive control and standard learner control. For example, based on the learner's performance on practice tasks or problems, the program can direct the learner to different learning modules that represent different levels of complexity, or to practice tasks that offer different degrees of instructional support. However, once this decision has been made by the program, the control reverts to the learner, allowing the learner to choose which kinds of practice tasks to perform from among an array of tasks characterized by different contexts and features tailored to the level of complexity to which the learner was assigned. Although there is still little evidence for the effectiveness of these shared control systems, and they would, as with guidance systems, require a substantial investment in resources for their development, the ability to characterize the learner and customize a set of practice tasks based on that characterization should benefit older learners more than conventional learner control systems.

10.2.6 Simulation and games in e-learning systems

Much has been made of younger generations of people who, by virtue of having grown up interacting extensively with digital games, may think, process information, and prefer to learn in different environments, and particularly highly interactive multimedia-infused learning environments that are relatively fast-paced, and emphasize high learner control and highly visual settings that encourage learning by exploration. We are certainly witnessing much greater reliance on high-fidelity simulations (Chapter 5) in some sectors, for example, in the medical industry where the benefits, in terms of reducing medical error related to risky procedures and learning procedures such as laparoscopy and robotics, are very clear. In fact, there is evidence that younger individuals who grew up playing video games are more proficient than older physicians at learning procedures such as laparoscopic surgery that involve visuospatial interactivity analogous to many digital games. With a report in 2006 indicating that 25% of game players are over the age of 50 (Clark and Mayer, 2008), perhaps there is potential for using simulations and games as vehicles for learning not only for younger adults, but even for older learners. However, although there are some guidelines related to the use of these approaches for learning, caution needs to be exercised in their application to older adult training and instruction.

The use and benefits of simulation for training have been noted in several places in this book (e.g., Chapters 5, 7, and 8). Basically, simulations are models of real-world systems. They capture the dynamics of these systems in ways that enable learners to realistically explore how their various interactions with the system affect system responses. In this book we have differentiated between high-fidelity physical simulators, for example, a mannequin that may be used to help train anesthesiologists in intubation procedures, and software simulations that essentially provide digital representations of a system that enables various types of interactions with the system to be explored within a conventional human–computer interactive environment.

From a training perspective, Clark and Mayer (2008) have made a distinction between operational and conceptual types of simulations. Whereas *operational simulations* are designed primarily to teach procedural skills (e.g., what to do when wind shear is encountered or a broken pipe is discovered in an industrial reactor vessel), *conceptual simulations* focus on learning strategic knowledge and skills to build far-transfer and problem-solving skills in specific domains. In the educational domain, conceptual simulations are designed to help students understand principles in domains such as physics or genetics, whereas in workforce training environments these simulations are designed to teach skills such as those that might be required in management, communication, and troubleshooting.

A distinction must be made between simulations and games. The essential elements that comprise games are: a well-defined goal that one or more players are challenged to achieve, the rules and constraints that govern the game, and a specific context that defines the game. Many games are instantiated through simulations, but some are not (e.g., game show quiz games). Naturally, many simulations designed for training purposes are not based on game formats. Rather, they emphasize human inputs to selected scenarios with the idea of providing the learner with realistic scenarios in which relationships between inputs and system outputs can be explored and comprehended.

10.2.6.1 Guidelines for designing simulations and games in e-learning

The goal in training and instruction is clear: to enhance learning. There is the danger, however, that incorporating a game-type environment within a simulation, for example, by having learners earn points for achieving certain goals, may also inhibit learning. This might occur through redirection of cognitive processing away from *essential cognitive load* (Chapter 5)—that part of the learning content needed for learning—for instance, when the game goals are not aligned with the instructional goals. Older learners, for whom it is likely to be more cognitively taxing to extract domain- or context-specific information relevant to the underlying principles or procedures that are being taught, are thus also likely to be more vulnerable to designs where the nature of the game goals may prove antagonistic to the instructional goals. Obviously, gaming features, which include ensuring a sufficient challenge (not too easy or too hard), providing adequate controls for the user to affect game outcomes, and instilling an element of curiosity regarding why exploratory activities lead to unpredictable outcomes, can make game-based learning more enjoyable. However, and especially with older learners, these fun features should not be incorporated at the expense of potentially undermining the learning objectives. Also, instructional designers need to keep in mind that what is perceived as fun is highly variable across individuals as well as across age groups.

In this regard, an important guideline that can help designers avoid this pitfall is to give consideration to matching the type of game with the nature of the learning goals. So-called "twitch" features that require learners to provide quick and precise movements in response to a displayed condition are a staple of many games but are inappropriate for older learners, and generally distract learners from focusing on deeper underlying principles that are being taught. A game designed in the style of an adventure, for example, that asks the learner to find everything a service representative did incorrectly in a simulated interaction with a customer, or all the reasons why pressure may have dropped in a reactor vessel, are better suited to training on hypothesis testing, troubleshooting, and problem solving. Games designed in the style of the *Jeopardy!* game show are more conducive to learning about facts, categorization of information, and concrete concepts. Although there is still much we do not know about matching game types to learning outcomes, this is a critical issue that needs to be carefully thought out, as older adults are less likely to be capable of compensating for inappropriate matching.

Even if games are designed to match the learning objectives, there are many other game design features that can lead to poor learning outcomes. Toward this end, Clark and Mayer (2008) offered a number of guidelines, some of which are fundamental to methods of learning in general (Chapters 7 and 8). The key principle in designing simulations or games as vehicles for promoting learning is to ensure that they do not overload WM while, at the same time, promoting essential or generative cognitive processing. The emphasis should not be on hands-on interactive activity directed at discovering important information that exemplifies highly exploratory environments, even though interaction can in certain circumstances contribute to meaningful learning (Section 10.1.5). Instead, the focus should be on the nature of the cognitive activities that are promoted within the learner.

To achieve this goal, it is critical that *guidance* be built into the simulation or game. This could be achieved, for example, by providing feedback to learners in response to their interactions with the simulation or game program that informs them about what has occurred or what actions they may need to take (see Section 10.2.10). Because it may be cognitively effortful to achieve the goals in games or simulations while absorbing important information needed for learning, the learner may place priority in meeting game goals, thereby compromising learning. This is what makes guided feedback, which can be provided either in the form of explanatory feedback (as opposed to corrective-only feedback) or as hints appearing between simulation trials, so critical. This type of feedback should not detract from the enjoyment aspects of the game. Also, it is important to be wary of emphasizing getting a high game score. For an older learner, a high game score may lead to increased game self-efficacy, but high game scores may not lead to better learning (i.e., to knowledge and skills that can be verbally articulated). A high score could reflect learning to play the game effectively, due to some form of implicit knowledge that is being accumulated, but if this knowledge does not translate into verbal knowledge, knowledge that can be thought about in WM, reflected on, and ultimately coded in LTM, then true learning was not accomplished (Chapter 7). This is another reason why instructional explanations during the course of the game or simulation that explain the principles or concepts that are being illustrated in these instructional environments are important; they bridge the gap that may exist between game or simulation performance and knowledge.

Another important guideline is to foster the opportunity for reflection on correct responses that were given in the game or simulation. Because these learning environments may be perceived as somewhat hectic in their execution, designers may be inclined to ignore this guideline as it tends to slow things down. However, with older adults especially, the opportunity to reflect on correct responses allows for deeper learning about the instructional content, and for maintenance rehearsal that makes transfer of knowledge and skills to LTM more reliable (Chapter 7). Reflection on incorrect responses should be avoided, as this approach has not been found to be effective and would likely induce cognitive overload in many older learners.

There are many more general instructional guidelines related to managing potential cognitive overload that should also be incorporated into learning systems which are based on the use of simulations and games. For example, a simulation or game should always begin with a relatively simple task or goal. This would allow the older user not only to gain confidence, but also enable extra cognitive processing to be available for managing any complexity that might still be arising from learning some of the features of the system. The progression to more complex tasks can, as discussed earlier, be guided based on dynamic adaptive control, for example, where the system infers that the learner has sufficient mastery of the topic to move on to a more challenging task, or it can be under the learner's control. If the simulation or game is highly complex in its functionality, consider turning some of the features off until the learner has mastered the most fundamental features, which is often referred to as a *training wheels* approach.

The use of fading or scaffolding (as discussed in Chapter 8 and in Section 10.2.10) refers to instruction that begins with a complete demonstration of a whole task or problem, and then gradually allows one to assume more of the responsibility for performing the tasks until practice problems are performed completely on one's own. For older adults, the use of a *pedagogical agent* that demonstrates at the outset how the game is played or how to interact with the simulation is very important, even prior to applying such a pedagogical approach to the transition from worked examples to practice through fading. This type of instructional support is critical for older learners as mastering the interface mechanics can be mentally challenging. These types of features are typically absent in gaming environments, where designers presume that players prefer to figure out the mechanics through exploration.

Because the use of games or simulations for instructing older learners is most likely to be for teaching facts, rules, concepts, and principles, rather than tasks that rely on speed, it is important that the pacing of games and simulations be under the control of the learner. To help the older learner cope with memory load, records of actions taken over the course of a simulation or game or other relevant data should be easily accessible by the learner so that the learner can reason about the instructional material based on a series of games or simulation trials. For example, in a simulation of how a financial advisor should negotiate the queries of a potential client, the learner should be able to retrieve a record that illustrates the problem contexts in which certain investment strategies were not recommended and others were.

Finally, the visuospatial challenges of game and simulation learning environments should not be underestimated when older learners are considered (see Section 10.2.8). Important information needs to be made perceptually salient, as it is more difficult for the older adult to process dynamic images quickly enough due to declines in processing speed (Chapter 7). Otherwise, the older adult may be left frequently asking him- or herself, "What was that I just saw?"

10.2.7   e-Learning as part of an embedded computer support system

In some work environments, an e-learning system component may be part of a larger computer-based system that supports work performance as well as instruction. Wickens et al. (2004) described such a computer-based system whose content could be used to support the performance of intermediate maintenance technicians as well as provide instruction to these workers. This system could consist of a computer-based integration of one or more of the following: guidance, advice, and assistance; data and images; and learning experiences and simulations. These types of support systems are sometimes referred to as *embedded computer support systems* as they may combine a number of common technologies that are often implemented on a more stand-alone basis, such as information databases, expert systems, help systems, adaptive aiding, and computer-based training. For example, a computer performance support system for intermediate maintenance technicians may combine: a job aid of procedures and descriptions of equipment repair, maintenance, and troubleshooting (that contains flowcharts, a technical manual, glossary, and videos of procedures); a training system component; and an illustrated parts breakdown (computer-aided design drawings of assemblies and subassemblies).

For older workers in particular, there are numerous ways that such a system, which integrates a variety of different computer-based tools to achieve diverse goals, could be overwhelming to use. In applying some of the methods of training discussed in Chapter 8, some of the challenges would be to ensure that the user understands the different modules or stand-alone subsystems that comprise the overall system, and specifically the purpose that each one serves; can identify each of these systems on the interface; can activate each of these systems; can navigate between each of these systems; and can enable each of these systems to remain active simultaneously in case there is a need for their integration. These concerns encompass many human–computer interface issues related to human information-processing tendencies and capabilities, as discussed in Chapter 7, such as perception (labeling, cueing, organization of information, clutter, balancing focused and selective attention) and cognition (awareness of what mode one is in and being able to interpret how actions translate into computer displays and the satisfaction of goals).

Once these higher-order design considerations are addressed, the individual computer-based technologies that comprise this system can be examined in detail to determine if they meet the relevant instructional guidelines, for example, guidelines related to worked examples and practice on the learning module components of the system (as discussed below) and guidelines related to multimedia for the multimodal aspects of the support system. With older adults, many other considerations come into play with such complex interfaces, such as fatigue, anxiety, and motivation (Chapter 6). This suggests the use of simple holistic tasks (Chapter 8) during initial practice to illustrate the integrated use of this system as a basis for instilling confidence and demonstrating meaningful benefits.

With such complex computer-based systems that may embed a variety of technologies related to performance support, learning, and refresher training, it is advisable during development and testing of these systems, especially if they are to be targeted toward older adults, to apply task analysis methods such as hierarchical task analysis (Chapter 8) as a basis for anticipating the different kinds of problems that the user may face when using the system to perform such tasks. As discussed in a tutorial entitled *Task Analysis and Error Prediction* in Fisk et al. (2009), this method, when combined with other task analysis related tools, can serve as a cornerstone for revealing errors and assorted difficulties that people, and especially older users, may encounter when interacting with a product or system.

10.2.8   *Visuospatial cognition and cognitive load in e-learning environments*

One of the factors in e-learning that truly differentiates it from many traditional learning environments that rely on text-based or printed material presentations of instruction is the enormous flexibility in visualization that can be afforded to the learner. Whether the instructional materials are presented in a computer-based (e.g., CD-ROM or DVD) or web-based environment, the instructional designer can consider the use of graphs, charts, photos, and video recordings. Furthermore, it is possible for many materials to be presented in static, animated, or interactive formats that can be moved, resized, hidden, or removed by the learner.

Despite the intuitive appeal of providing such flexibility to the learner (at least in the visualization media component), demonstrating that enhanced visualization promotes learning is not so simple. The implication from the broad discussion of older adult information-processing capabilities laid out in Chapters 4 and 7, and the integrated human information-processing model (Figure 7.4) that more clearly highlights the ways in which cognitive load can derive from visualization demands, should certainly alert instructional designers to the risks to older learners of presenting visuospatial materials.

This is not meant to imply that the array of creative visualization artifacts and methods that have been conceived for instruction have not been found to be advantageous. Undoubtedly, given the esoteric nature of the underlying instructional material in many topics (especially in mathematics and science) and the limitations in relying on purely verbal models for comprehension, many difficult concepts are likely to be better grasped with a corresponding and integrated visualization component. However, when instructing older adults on cognitive tasks, the maxim, "Less is better," may indeed apply. The key, as always, is to balance the load imposed by the visualization artifacts (e.g., as might derive from the type of visualization and the degree of control over, and interactivity with that artifact) with methods of instruction that enable relevant, retainable, and generalizable knowledge and skill to be systematically acquired.

The argument for presenting information both through visuospatial and verbal "codes" (Chapter 7) is central to appreciating the potential benefits of multimedia models of instruction (Figures 7.2–7.4). For example, these models not only enable cognitive load to be distributed over the two primary WM subsystems (Chapter 7), but also over the visual and auditory modalities when instruction incorporates narration. Multimedia also can facilitate a stronger memory trace in LTM through the integration of verbal and pictorial models by enabling a greater range of cues, deriving from pictorial as well as verbal information, to trigger associative memory recall.

Another advantage of pictorial visualizations during the course of learning is that they provide an external representation of information containing analogical properties. As noted in Chapter 7, having available *knowledge in the world* as opposed to *knowledge in the head* can reduce cognitive load during task performance. Similarly, during learning, such external representations could free cognitive resources from the task of constructing or maintaining mental models based purely on verbal information by providing a pictorial reference that the learner can repeatedly revisit as part of the learning process.

The possibility also exists that such visualizations may provide the opportunity for deeper processing of the learning material, especially if interactive tools are provided to the learner. For example, such tools may allow for situating graphs and pictures side by side to enable comparisons; changing colors to enhance the perception of objects; or for speeding up or slowing down animations or videos to make the information being presented more commensurate with one's cognitive processing capacity or, through fast play, highlight how a process occurs by emphasizing change over time. In fact, any procedural task whose sequential or temporal relations are intrinsic to its performance can benefit from the flexibility afforded by computer-based visualizations. Of particular relevance to instruction targeting older adults is that such computer-based visualizations can simplify the comprehension of complex material, especially when the underlying concepts can be characterized by cause-and-effect relationships (Oestermeier and Hesse, 2000). For example, such visualizations could be useful when trying to explain the difference between a gas stovetop oven and a ceramic stovetop oven with respect to how a change in temperature translates to a pot of water boiling. As was noted, not only comprehension, but also memory of this (e.g., cause-and-effect or sequential) information can be enhanced through pictorial visualizations.

However, as repeatedly emphasized in this chapter, particularly with regard to multimedia as well as throughout this book, designing instruction for older adults generally involves tradeoffs. Specifically, these tradeoffs must take into account the nature of the to-be-learned material, the characteristics of the learner, and the design formats and tools presented or available to the learner. For example, in the discussion on multimedia, the possible benefits for older adults (and even other learners) of static diagrams over animation was noted. Thus, if creative ways could be found for demonstrating cause-and-effect, temporal, or sequential relationships using sequences of static snapshots that the learner can easily control, these may be more suitable for older learners. Of course, the possibility would still exist for enabling the older learner to view an animated or fast video presentation that can provide a more compelling presentation of these relationships. However, these options should probably be deferred until sufficient learning has been established and the risks to learning resulting from viewing such presentations are reduced. A basis behind these and similar recommendations is that perceiving and comprehending complex computer-based visualizations can be cognitively demanding.

A number of the guidelines related to design of multimedia instruction (Table 10.2) are intended to circumvent the constraints associated with the human information-processing system for older adults. However, as emphasized in Chapter 7 we still know very little about how the demands associated with the executive control mechanism in WM, which is critical to integrating verbal and pictorial information, may undermine multimedia learning approaches. Also, for older learners, the availability of flexibility in visualizations must also be viewed with caution. For example, requiring the learner to perform relatively complex transformations of the visually presented material, such as rotating or aligning information, or more generally, any requirement for the visualization material to be mentally transformed in some manner, may be too cognitively demanding for many older adults.

A second reason for exercising caution in designing visualizations in e-learning instructional environments for all learners, and particularly older learners, is individual differences in visuospatial abilities. People with high visuospatial abilities are more likely to benefit from pictorial-type visualizations in learning or instructional scenarios (Gyselinck et al., 2002). Older learners with low visuospatial ability are likely to be especially disadvantaged in using such visualizations as they are already more prone to exhibit declines in spatial memory ability. However, low spatial ability older adults may actually be able to derive learning benefits that would have otherwise been difficult for them to obtain. For example, this might occur if the pictorial visualizations are designed to impose low cognitive loads and are easily integrated into the verbal models of the subject matter. If the learning topic naturally benefits from such integration, such older learners may thus be able to achieve deeper learning or better comprehension than they would have otherwise been able to achieve.

In summary, with increasing technological sophistication of computer-based visualizations that can be embedded into e-learning environments, designers need to exercise caution when the instructional material may be targeting older adults. Careful consideration of the nature of the to-be-learned material can go a long way toward ensuring that the pictorial representations offer the potential for more efficient and improved comprehension and, at the least, do not prove to be a liability to the instructional process.

10.2.9   Learning "alone": Asynchronous e-learning and older adults

The ability to effectively use e-learning methods such as CD-ROMs and online training platforms is critical for older adults as this type of learning environment is becoming very common in many organizations as the means for imparting skills to employees. The inability for older adults to use e-learning platforms thus could result in their exclusion from employment opportunities, including being rehired by their former employers or the opportunity to seek part-time employment, as well as the ability to engage in other types of endeavors. However, very little is known about older adult e-learning capabilities or attitudes.

The pilot research of Stoltz-Loike, Morrell, and Loike (2005) represents one of the few studies in this area. Their first pilot study, which encompassed a small sample of older adults, investigated the ability for these adults to learn skills related to career management and pre-retirement planning using a subset of the *BusinessThinking*™ courses in a CD-ROM format. Overall, the participants found the information presentation and design of the software to be good, with the highest ratings given to the software's navigation system. Based on pre–post tests, the findings also indicated that the e-learning course resulted in improved knowledge about the topics.

In their second pilot study, these researchers investigated whether *BusinessThinking's* CD-ROM e-learning platform could be used to teach older adults how to use a technology application, specifically, Microsoft PowerPoint. Most of this sample, although they had some computer experience, had no prior experience with this application and those who did rated themselves as having low skills. The participants were required to complete 18 basic PowerPoint tasks prior to the course and then again one week after the e-learning course. As with the first pilot study, ratings related to the design of the course were relatively positive. Interestingly, and in line with the *passive versus active* debate regarding learning, the investigators of this study attributed the much more substantial improvement that was found between pre- and post-test scores exhibited in the PowerPoint instructional scenario to the fact that the participants had to perform—that is, practice—hands-on manipulation of computer procedures in order to learn them, thereby reinforcing memory of the material. In contrast, in the first study the requirements were only for remembering the course material text.

These findings are encouraging in that the courses were not customized for these samples of older adults, suggesting that e-learning platforms might have an even better chance of success if they were specifically tailored for older adults and also afforded certain degrees of individual customization. In this regard, these investigators caution that many of the available e-learning courseware materials assume a relatively sophisticated knowledge of technology and "e-terminology," and familiarity with use of e-learning courseware. This may nicely accommodate technologically savvy people, however, those unfamiliar with computer-based training, which would include many older adults, are likely to find such learning platforms inaccessible and thus unhelpful.

Because e-learning platforms are typically designed to be used at the learner's own convenience, often independently, the learner's self-efficacy, especially as it pertains to computer technology, is potentially a significant issue (Chapters 4 and 6). Many older learners with low computer self-efficacy may encounter difficulties in e-learning environments that could engender frustration, reinforcing already low levels of self-efficacy, and which could lead to abandonment of the e-learning platform. In general, expectations of lower self-efficacy related to technology interaction among older learners and their own metacognition of declining cognitive abilities is consistent with evidence that older adults benefit more than younger adults from face-to-face instruction (Sullivan and Duplaga, 1997), perhaps in part because they are more likely to ask for help.

A challenge, then, is to construct e-learning platforms that can adapt to the older learner's level of technology savvy (Section 10.2.5). If it is determined that the older learner is experiencing difficulty with the interactive aspects of the platform, it is necessary to identify the nature of those issues and ensure they are overcome prior to engaging the user further in learning activities. One way this can be done is to have a preliminary module that attempts to gauge the user's degree of knowledge regarding computer-interface tools and e-terminology, and correspondingly to make available alternate versions of the course that best fit different learner needs for user-interface design friendliness.

In addition, the opportunity for users to easily broadcast their concerns or "speak to" the e-learning platform would enable issues and points of confusion to be potentially addressed and rectified almost immediately. This, in turn, could provide the confidence that the older learner may need to move ahead with the course materials. These additional e-learning design considerations obviously add time and thus cost to the development of the platform, but are well worth the effort if the intention is to extend the potential benefits of the learning platform to the older adult population.

10.2.10   Worked examples in e-learning environments

e-Learning environments that embrace multimedia provide excellent opportunities for incorporating worked examples into a training or instructional program. Worked examples, which can be viewed as a method of instruction, are a step-by-step demonstration of how to perform a task or solve a problem. A number of issues associated with worked examples, and in particular how they should be sequenced, were presented in Chapter 8. Here, we revisit this method, but within the context of e-learning systems.

It is not uncommon for people to bypass text-based descriptions of how to do something or information concerning a learning topic for an available worked-out example. For example, rather than read, line by line, about how to use a spreadsheet, how to appropriately sterilize an instrument, or how to conduct a commercial loan negotiation, it would seem more efficacious to directly consult a worked example that succinctly lays out what needs to be done. The reason why most people would opt for the latter is that it minimizes cognitive effort (Chapter 7). Early in learning, as people build new knowledge, they need to conserve their limited WM resources. Reading and keeping track of the information laid out in text and following that experience with a single worked example and

ensuing practice on problems, which is a very traditional way of providing instruction, consumes a great deal of WM capacity. Initially placing
more reliance on worked examples can free WM capacity so that it can be
directed at reflecting on the material, that is, on learning. At some point,
as emphasized in Chapter 8 with regard to sequencing worked examples,
the shift to practice and problem solving has its benefits as it helps to automate the new knowledge that has been accumulated.

With older adults, there is an even greater risk that if cognitive load is
not managed effectively during the initial stages of instruction, very little
if any new knowledge will be obtained, and a cycle of frustration, reduced
self-efficacy, and loss of interest may set in. Worked examples are thus an
option that should be strongly considered with older adults. A number of
existing principles can be used to guide the design of worked examples
(Clark and Mayer, 2008), and these principles are, to a large degree, also
relevant for older learners. One principle, which is consistent with the 4C/
ID model (Chapter 8), is the use of transitioning from worked examples
to full problems using scaffolding, or what is also referred to as a *fading* approach. With gradual fading, the initial worked example is fully
worked out. With subsequent presentations of worked examples fewer
and fewer steps comprising the solution of the example are worked out,
and the burden is gradually shifted to the learner to complete more of the
example's steps until the learner completes the problem on his or her own.

One concern with this method is the possibility for overdoing the presentation of worked examples so that studying the worked example provides no new learning. In fact, according to some, it may actually impede
learning as it would have been more effective to resort to practice in order
to automate the new knowledge and skills that were acquired. With older
learners, however, we suggest erring on the side of more worked examples
rather than fewer prior to transitioning to practice exercises, as establishing resilient memory traces may prove more difficult for older learners.
Thus, it is probably safer to allow the older learner more opportunity to
have WM capacity available to ensure new material is being processed
prior to automating that learning.

A second principle calls for including self-explanation questions with
the worked-out steps of the worked examples. As an illustration of this
principle, consider a worked-out example that provides, in a stepwise fashion, the procedure for performing an emergency shutdown operation in
a plant. Following the worked-out solution to a step in this procedure the
program may query the learner to choose, from four choices, the appropriate explanation for that solution step (e.g., by closing the valve, contam-
ination of the product will be avoided). The inclusion of self-explanation
questions promotes interactivity and thus avoids the possible tendency
with worked examples, due to their inherent passive nature, of ignoring

them. It also promotes more meaningful learning as it encourages the learner to think about the material.

A third principle recommends supplementing worked examples with explanations. This is especially important with older learners as it should not be assumed that they understand the solution steps that are shown. If in fact they do not, the expectation that they have to respond to self-explanation questions may engender increased frustration. Many older learners would therefore probably benefit from detailed explanations, at least of the initial worked examples, that provide the underlying rationale to the solutions of the worked-out steps. The explanations should not be too long or complex; that is, designers should guard against the need for excessive cognitive effort to process the explanations. Consistent with multimedia guidelines, these explanations should be positioned in close spatial proximity to the worked-out step, and this same principle would also apply to the presentation of self-explanation questions. As the older learner gains expertise, the presentation of explanations could be made optional or in response to an error that the learner makes on a step. In this way, the older learner can gain confidence through the need for less dependence on explanations except when an error occurs or when help is requested.

Another important principle concerns the type of learning that worked examples are intended to support. In this regard, two types of training goals can be distinguished: *near transfer* and *far transfer*. Near transfer generally pertains to instruction on procedures or tasks that are performed approximately the same way each time. The key in designing worked examples to support near transfer is to emulate, in the worked examples, the work (or other environment) that is the target setting for the task or procedure. Capturing the context of the setting is critical, as it is important that the same cues that signal the steps that need to be taken in the task environment are presented to the learner for encoding in the training environment. For this reason, worked examples designed to support near-transfer learning are often in the form of demonstrations. As stated and implied in various parts of this book, older adults do less well in dealing with gaps in information needed for understanding a learning topic or to perform a procedure. Therefore, dissimilar contexts are more likely to generate confusion for older learners. This, in turn, translates into increased cognitive load stemming from attempts by the learner to reconcile the differences that existed or were perceived to exist between the training and task performance environments.

In contrast to near transfer, far transfer in learning relates to training whose goal is to develop skills that enable the learner to adapt the learning material to new situations. This applies to many work situations, such as sales, financial advisement, and many maintenance troubleshooting tasks, as well as many tasks performed in the medical and legal work domains. In these situations, being trained on an invariant set of steps is not purposeful with regard to managing the less predictable situations that may lie ahead. Instead, the goal is to provide a "core template" of learning material that can enable the learner to adapt the training to solve new problems. These problems may appear to the learner to be different from those that were experienced during training but, in fact, are based on underlying principles that governed the training.

A key guideline in designing worked examples for far transfer learning is to provide examples possessing different contexts or story lines that, when taken collectively, can induce the kind of flexibility in thinking that is needed for adapting to different problem situations. The idea of presenting varied contexts in simulations is, in fact, a key concept underlying transfer of training as discussed in Chapter 5. However, it is not simply the act of presenting varied-context worked examples but, in addition, the manner in which learners become engaged with these examples that can promote adaptive or strategic knowledge. Thus, another guideline is to incorporate at least two worked examples into the training in which the contexts or story lines are varied, but which embody the same underlying principles capable of supporting far transfer. The training should also engage learners by encouraging them to actively make comparisons between the examples. For example, the initial worked example can be accompanied by self-explanation questions, as recommended in principle two. The second worked example can then be displayed on the screen alongside a summary of the solution to the first worked example. Promoting an active comparison by the learner of the two worked examples can be accomplished by asking the learner to identify the underlying principles that are common to both examples. As implied in Chapters 4 and 7, this approach provides the basis for developing the knowledge and skills essential for generalization, which is a fundamental aspect of expertise.

Finally, a sixth principle associated with worked examples calls for incorporating relevant multimedia guidelines as presented in Table 10.2. For example, when presenting the steps of a worked-out example, rather than using text alone consider presenting relevant visual information for illustrating the steps. However, the spatial contiguity principle should be applied by placing the text, numbered according to step, in close proximity to the relevant visual information (Figure 10.4). Also, if there are many steps in the worked example, the steps should be grouped into meaningful chunks. This will help reduce WM load associated with processing the example, as well as encourage building schemas in LTM whereby several steps may become associated with an underlying principle of the training (Chapter 7).

The use of audio is generally recommended, especially for accompanying visuals. However, with worked examples incorporation of the audio modality in the form of narration could actually impose additional

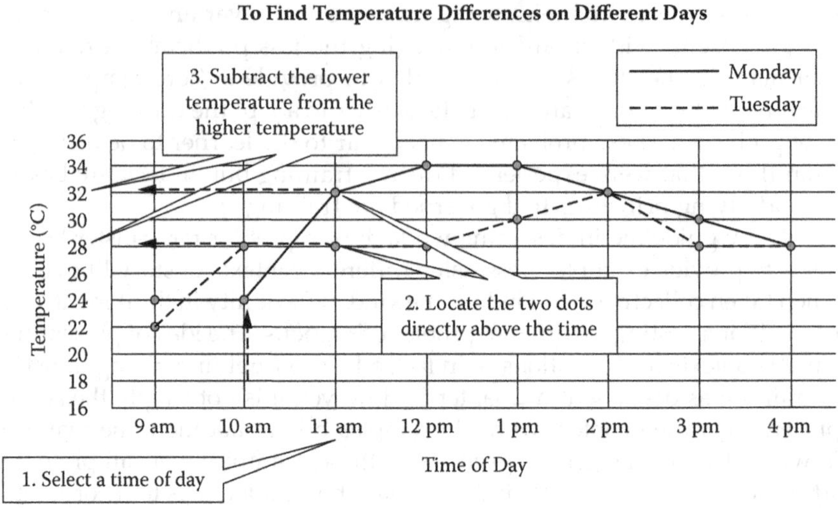

To Find Temperature Differences on Different Days

Figure 10.4 An illustration of the integration of worked example steps into a visual. (From Clark and Mayer, 2008. *E-Learning and the Science of Instruction: Proven Guidelines for Consumers and Designers of Multimedia Learning*, 2nd ed. San Francisco: John Wiley & Sons. With permission.)

cognitive load on the learner. This is because some worked examples rely on fading, where the learner must complete some of the steps. However, all the steps would need to be presented in text, as this would enable learners to review those steps at their own pace in order to complete the steps that are faded. The inclusion of self-explanations in worked examples also benefits from steps being presented in text to enable flexible review of those steps for the purpose of identifying the appropriate learning principles. Unless the use of audio can provide a unique benefit associated with the visual, for older learners it is probably best to use a combination of visuals and appropriately integrated text in the presentation, and always allow the learner complete control of the pacing through the examples and corresponding succession of screens.

10.2.11   Practice in e-learning environments

As emphasized in Chapters 4 and 5, practice is the basis for acquiring skills in task performance and problem solving. Thus, not only is it important for instructional programs, including e-learning programs, to present worked examples effectively, but careful consideration must also be given to how practice is managed as the learner is transitioned from worked examples to full practice tasks or problems. This transitioning is especially critical for older learners, as engaging in full practice problems can be extremely mentally taxing for them. The gradual imposition of mental load through scaffolding (Chapter 8) or fading of worked examples is therefore critical to ensure that the combination of worked examples and practice leads to the ability to select, organize, and integrate the learning materials, and to retrieve this new information, in order to foster efficient learning.

For almost all types of learning situations, a longstanding precept in human learning is that *distributed practice* (Chapter 4), whereby practice problems are distributed or spaced over time, provides better learning in terms of long-term retention of the material than *massed practice*, which imposes practice on the learner in a more condensed time period. The benefits of distributed practice are, however, not realized immediately and thus may not be observable, which speaks to the need for long-term evaluation of training and instructional programs (Chapter 11). As recommended in previous chapters, distributed practice, both within and across practice sessions, should be administered to older adults. Unfortunately, some older adults may be more susceptible to the practical constraints implicit in distributed practice; obviously, the longer time periods, both within a training session and across sessions, would require a greater commitment. This is another reason why it is so critical that older adults are sufficiently motivated to pursue the instructional program (Chapter 6).

For practice on tasks or problems that people will perform in job situations, it is important that practice is tailored to the nature of the performance requirements. First, and as has been repeatedly emphasized, a task analysis is needed to specify the physical and cognitive requirements of the task as it will be performed in the job environment. This is needed to ensure that the features or cues intrinsic to the work context are encoded during the interactions the learner has with the practice assignments so that the correct cues will be available in LTM when transfer of training to the work situation is required. Establishing this clear connection between features common to the practice and work environments is very critical for older learners, as they are generally more susceptible to confusion deriving from inconsistent mappings between these environments.

It is also important to acknowledge that tasks in work settings may differ on a number of characteristics that could affect the way practice should be designed, and some of these characteristic differences lend themselves quite well to the flexibility that e-learning can afford to the design of practice. For example, some jobs require that skills be practiced to the point of automaticity (Chapters 4 and 7) prior to the first time such tasks are performed in the work environment. This is often the case when task performance has safety or high economic risk implications (e.g., where an inappropriate shutdown of a production process due to poor recognition or understanding of process variables can result in millions of dollars in losses). In other cases, automaticity in task performance may be desired but may not be needed as an outcome of practice; instead, this degree of skill can be obtained while performing the tasks on the job. In other cases, for example, where the emphasis in training is on discerning and resolving different kinds of issues customers may have concerning purchasing a financial product, the emphasis in practice should be on comprehension of underlying concepts. In these cases, it is the nature, rather than the degree of practice that is critical. Still other job situations may require different degrees of mixtures of automaticity, concept understanding, context recognition, and reflection.

Obviously, the amount of practice should be adjusted to the job environment requirements, and e-learning systems, given their flexibility, offer an efficient means for realizing such goals. e-Learning systems can also tailor the nature of the feedback given during practice to the job requirements. As with principles related to feedback with worked examples, all learners, and especially older adults, will benefit from feedback related to the correctness or quality of their responses. However, simply providing nonspecific feedback (i.e., informing learners whether their responses were correct) should be avoided. Instead, learners should be given *explanatory feedback* that is intended to circumvent the building of incomplete or incorrect mental models (Chapter 7), and that helps close gaps in understanding of concepts or guides them (e.g., through a brief demonstration) with regard to the correct approach or procedure. Unfortunately, e-learning tools make it very easy for designers to implement corrective feedback. In the interest of efficiency or pressure to roll out an instructional program, designers are likely to avoid the cognitive effort needed for crafting the more intensive explanatory feedback systems. This tendency needs to be avoided when older learners are considered, as simply automating a corrective feedback approach into the e-learning system can not only disrupt learning in this population but also, perhaps in part due to lower self-efficacy, induce greater frustration.

Finally, principles of multimedia design that are relevant to design of worked examples will be relevant as well to the design of practice assignments. For example, the correct answers along with the corresponding explanatory feedback should be positioned in close spatial proximity to the questions and to the responses that were given by the learner. Also, learners should be given the option to review explanatory feedback at their own pace and the opportunity to return to earlier points in the practice session. Although narration offers a powerful way, in multimedia systems, of diminishing the cognitive load on the learner, practice questions, feedback, or onscreen text directions should not be narrated (Clark and Mayer, 2008). Extraneous load should also be minimized by avoiding the temptation to include music, sounds, pictorials, stories, or other artifacts that may have entertainment value but, especially for older learners, are distracting and remove cognitive resources that are needed for essential learning.

10.3 Avatars and virtual worlds

Avatars are digital representations of people that can be used to facilitate social interactions and communication in Internet and other virtual environments. These multimedia agents can help to personalize the learning experience by creating an environment that more closely resembles human-to-human conversation and a sense of partnership with the instructional system (Table 10.2), thereby potentially inducing in the learner deeper cognitive processing. Their implementation within a virtual teach-back technique, as a computer alternative to the live teach-back technique often recommended for older persons, was discussed in Chapter 8.

Another application in which avatars may prove useful is in the provision of medication advice. A prototype spoken dialogue system referred to as Chester, described in Allen et al. (2006), was intended to enable automated in-home conversational assistants help patients manage their prescriptions. The assumption is that the traditional way in which this information is made available, which is usually either on the prescription information sheet or online, may be problematic (e.g., too difficult to read, access, or comprehend) for many patients such as older adults. Chester was not designed to make medical decisions, but rather to help patients follow instructions given to them by their doctors. One challenge with this system is that it must be very confident that it has understood the user's question in order that it can proceed in its dialogue without the benefit of obtaining clarification or confirmation from the user. Because conversational assistants are generally based on complex computer architectures that often involve the integration of many specialized modules, achieving such degrees of confidence may be very difficult. Thus, as a potential advisor or instructor, such systems may risk losing the trust of the user if an adequate level of reliability cannot be demonstrated. Older users in particular, as with any adaptive instructional system, may experience greater frustration, leading to disuse of the system.

In the prototype work of Allen et al. (2006), the medication advisor system was embodied as a graphical avatar in the shape of a pill (Figure 10.5) whose actions could convey various system states such as idle, attentive, thinking, and speaking. Although these avatar personifications were relatively simple, in addition to providing important information, they were capable of engaging the attention of the user (e.g., in conveying to the user that it needed extra processing time). Older people can especially benefit from the important cues that such natural gesturing can provide.

A demonstration of the dialogue between the conversational assistant and the patient is demonstrated below for the case of an older woman who is taking a small number of commonly prescribed medications and who wakes up with a headache and decides to consult Chester for help. The dialogue begins with the patient asking, "Can I take an aspirin?"

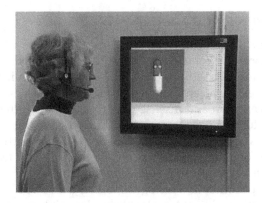

Figure 10.5 Screenshot of the Medication Advisor. (From Allen et al., 2006. *Journal of Biomedical Informatics*, 39: 500–513.)

Note that the assistant must identify that the patient is not asking about whether they could physically swallow an aspirin, but rather whether it is advisable to take an aspirin given their medical condition. This requires recognizing the intention of the user. The complete dialogue is as follows (Allen et al., 2006, p. 504):

Patient: Can I take an aspirin?
Chester: No, you are taking Celebrex and Celebrex interacts with aspirin.
Patient: Oh. Could Celebrex cause headaches?
Chester: No, headaches are not an expected side-effect of Celebrex.
Chester: Do you have a headache?
Patient: Yes.
Chester: How severe is your headache on a scale of one to five?
Patient: One.
Chester: Your doctor says you can take Tylenol for mild headaches.
Patient: OK, I'll try that. Why am I taking Celebrex?
Chester: You are taking Celebrex to help with your arthritis.
Patient: Should I take one now?
Chester: Yes, you are taking Celebrex every morning.
Patient: Is there anything else I need to take now?
Chester: Yes, you should take your Prinivil.
Patient: Which one is that?
Chester: It's the yellow one [shows picture].
Patient: Could that be causing my headaches?
Chester: Yes, headaches are an expected side-effect of Prinivil.
Chester: Tylenol should help.
Patient: OK, thanks.
Chester: You are very welcome.

Another demonstration of the instructional use of avatars was in training physicians to perform home safety assessments for older patients (Andrade et al., in press), which represents an integral part of geriatrics fellowship training. Simulated home safety evaluations in 3D environments that enable the user to interact with the virtual environment through avatars may enable, in certain learners, and through the subjective sense of presence or being there, a cost-effective and safe method for acquiring the skills needed to detect home hazards. From a learning perspective, one advantage of the virtual world as compared to the physical world is its ability to emphasize hazards that may appear subtle in the home environment.

For this study, an interactive 3D virtual two-story New England style home was created in Second Life, an online virtual world environment created by Linden Lab. Second Life is one of a number of online virtual world environments in which users can interact with each other by creating avatars that inhabit the virtual worlds. For example, users acting through their avatars can participate in group activities, attend virtual classes, create and trade virtual real estate and services, design products, and engage in lifelong learning (Wankel and Kingsley, 2009).

The participants in this study were senior medical students, residents, and fellows who had previously conducted home safety assessments. Each participant was represented by a physician avatar that was navigated through an interactive 3D virtual house with the task of identifying as many safety hazards (derived from an inventory of 50 safety hazards) as they could (Figure 10.6). In addition to their performance, the participants' self-efficacy (confidence in one's skills at identifying home safety hazards), spatial ability, subjective mental workload, and presence

Figure 10.6 The living room of a two-story interactive 3D virtual house in Second Life. 1. Sofa set. 2. Lighted candles on table. 3. Tipped-over rocking chair (not displayed). 4. Lit unattended fireplace. 5. Furnace with smokestack. 6. Wall-mounted LCD TV with stand. 7. Stacks of newspapers. 8. Black thick carpet. 9. Ants on floor. (From Andrade et al., in press. *Gerontology & Geriatrics Education.* Available online: http://www.tandfonline.com/doi/full/10.1080/02701960.2011.611553. With permission.)

were assessed. Results from a usability questionnaire indicated that all participants perceived their task experience as positive, had very little to no difficulty navigating the avatar in the virtual world, and found the environment to be user friendly and conducive to learning.

The logical question that follows is whether older adults themselves can learn to identify hazards that might be present in their home environments by navigating such 3D virtual environments. Older adults may be forced to move into different living environments, and costs to perform such assessments in their homes by physicians could become prohibitive, shifting the onus to the older patient to make such assessments. In light of evidence that individuals with high and low spatial ability experience different degrees of success with interfaces that present information spatially and require navigation in a virtual world, there is always the concern that older adults with low spatial ability may be disadvantaged in these learning environments. Most likely, many older individuals may require extra practice and increased guidance in navigating the avatar in the 3D virtual world.

However, given evidence that a person's spatial navigation can be improved through practice in the virtual world (Rose et al., 2000), the possibility exists that with practice and guidance in the virtual world this type of training could provide the necessary memory cueing that would enable older adults to identify hazardous objects or conditions in their own living environments. What would be of interest is whether a heightened sense of presence for older individuals in the virtual world correlates with their ability to translate successfully or generalize practice on this task to real world performance. It may be that older adults require greater fidelity in the virtual simulated world—that is, a more concrete rather than general representation of the home environment—for facilitation of such memory cueing (Chapter 5).

In any case, there are many applications that one can conceive of where immersion in a virtual world through use of a personal avatar may be a cost-effective and easy way to provide instruction, especially when it involves identification and discrimination among various critical stimuli as a basis for decision making. Second Life and other 3D virtual worlds have been particularly useful in collaborative situations. These environments thus could be useful in training older people for tasks involving team performance, which may be demanding for some older workers (Smyer and Pitt-Catsouphes, 2009). Virtual worlds can provide models to older adults of what others in the team are doing, while also helping them to better understand the overall project goals and how to evaluate progress toward those goals. For such relatively complex instructional applications of virtual worlds, fundamental methods of instruction such as scaffolding (Chapter 8) could be incorporated, as well as emphasis on instruction on the basic skills of inhabiting virtual worlds such as Second Life.

The Second Life virtual world allows users to create avatars of any appearance, age, race, and gender, and there is evidence suggesting that pedagogical agents such as avatars may enhance engagement and learning beyond computer-mediated communication without such agents (Atkinson et al., 2005). Virtual worlds also offer possibilities for role-play that do not exist or are very difficult to generate in face-to-face training scenarios. For example, Second Life can allow an older person to play multiple roles in role-playing scenarios such as the role of financial advisor, customer, and appraiser (e.g., a manager). These role-playing scenarios can provide a mix of thinking (partly through enhanced self-awareness), emotion, and action that can strengthen learning as problems can be seen from multiple points of view. In addition, online role playing provides a unique benefit to older adults: whereas in real-life role playing an immediate interpretation of and reaction to signals from others is required, online role playing, in contrast, allows more time for reflection. This advantage could translate into more effective skill acquisition and retention (Chapters 4 and 5).

10.4   Recommendations

Throughout this chapter, many recommendations in the form of principles, guidelines, and suggestions were offered, including those that comprise Table 10.2. In this section, these recommendations are briefly summarized. Although the recommendations are grouped into multimedia and e-learning categories, these distinctions are often transparent as e-learning systems typically embrace multimedia formats.

10.4.1   Multimedia

- Include words (either printed text or spoken text) with graphics (charts, graphs, photos, static illustrations, videos, or animations) to make more effective use of limited working memory capacities and encourage deeper thinking about the material through connections that the learner can make between these two different types of representations. Graphics could be vital if the instructional content relates to phenomena that are difficult to visualize or imagine or relationships that could be difficult to contemplate.
- Do not overload the visual sensory channel by presenting too much text and graphics material. Consider using narration to present the text. If the material is relatively complex, segment the material through logical breaking points. Use a lead-in to the next segment that acts as a refresher for the previous segment(s) and maintains the continuity between segments.
- Avoid presenting information that is extraneous to learning as it can divert cognitive processing away from essential learning. Thus,

avoid background music, excessive sound effects, amusing graphics, or other types of information that are extraneous to the points being emphasized in the training or instruction.

- Cue the learner about how to select and organize material by stressing certain phrases in speech, adding headings, or highlighting certain images with arrows. However, do not use this strategy excessively to the point where this information constitutes extraneous load.
- Integrate critical text within the graphics to create spatial contiguity and thus reduce the overhead associated with back-and-forth sampling, scanning, and matching of text and pictorial information. If there is insufficient space to place the text within the graphics, avoid cluttering the pictorial elements and instead consider using narration concurrently with the pictorial presentation to capture this additional textual information.
- Avoid presentations of redundant narration and onscreen text if the presentation also contains graphic elements. Use narration alone for presenting the words accompanying the graphics. However, if the presentation does not involve animation or pictures, consider using concurrent narration and onscreen text (i.e., verbal redundancy) rather than narration only or onscreen text only.
- Ensure temporal contiguity by synchronizing in time the presentation of corresponding visual and auditory information (e.g., by having narration coincide with actions shown on videos). If synchronization cannot be accomplished, present very small segments that alternate between narration and the corresponding animation information to reduce the possibility for overload.
- Before using animations or videos in multimedia instruction, determine how essential these highly information-intensive presentations are for learning. Consider use of a series of static snapshots in place of dynamic animations. However, for training situations that involve procedural tasks (how to do something), use dynamic presentations such as animations, but ensure that the rate of presentation of dynamic media is not too rapid and is under the control of the learner.
- Always ensure that any prior knowledge needed for subsequent processing of learning materials has been sufficiently rehearsed so that it is likely to be retrieved from the learner's long-term memory. This is critical for guiding the process by which the learner will select, process, and integrate this subsequent information.
- At the beginning of an instructional session, use a narrator's personalized style as a calming influence to allay possible fears related to the impending instructional sessions. Throughout the instruction, personalize the learning experience by using a conversational approach that more closely resembles human-to-human conversation. Avoid formal speaker styles that simply focus on delivering

information. Try to generate a "social presence." Instead of a passive voice approach to instruction, use a friendlier second person approach with more informal language.

- Use narration to help direct the user's attention to visually presented information that may otherwise not get noticed through appropriate emphasis and directives provided in the narration.
- Avoid overloading the visual and auditory modalities when integrating narration and visually presented information. Consider pausing narration to give the learner the opportunity for processing what might be an abundant amount of screen information. Carefully think about the amount of narration to use at any given time depending on the amount and type of visual information that is being presented.
- Ensure that critical interactions with input devices and other auxiliary technologies or objects are mastered prior to initiating the learning sessions so that cognitive processing resources are not allocated to the use of any of these devices during the learning sessions. Carefully consider the design of interactive features.
- Ensure that the visual display is large, that text is large and pictorial information is clear, that onscreen windows are identifiable and clearly distinguishable, and that there is sufficient illumination and contrast between the screen elements and the background.
- Make certain that the older learner has a comfortable chair; that the volume of any narration or sounds is adequate; and that controls for manipulating the presentation are easy to use. Provide for adjustability in font size and loudness. Ensure that the computer system on which the multimedia presentation is being implemented is comfortable to interact with, attractive, and pleasant to use.
- Provide opportunities for the learner to pause the presentation and to easily continue the presentation following the pause. Following a pause, the learner should be able to identify previous sections of the presentation easily, go back to any of these sections, and go to any point forward as well.

10.4.2 e-Learning

- The balance between program control and learner control in an e-learning system should be in the direction of learner control with regard to the various control features that the e-learning system offers. That is, trust the judgment of (older) users in knowing what they know or do not know with respect to the learning session. However, it is essential that additional time be allocated for older learners to enable them to become familiar with the learner control features.
- To ensure that worked examples or practice exercises essential for acquisition of fundamental knowledge and skills are not

disregarded, consider placing these features under program control or at least make them difficult to be bypassed by the learner.

- Consider adaptive control features that can detect, perhaps based on some prior assessment of the learner, if the e-learning system contains technological features that could be troubling for the older user or terminology that may be unfamiliar. If so, provide the capability for the program to branch to an interface or lesson that is closer to the user's level of technological expertise or knowledge.

- Consider implementing adaptive dynamic control that enables the number and complexity of practice tasks to be adjusted, based on the learner's performance on practice tasks, in order to increase the efficiency of the learning sessions.

- Consider implementing adaptive guidance whereby, based on the learner's performance, the program may make suggestions. For example, the adaptive guidance feature may suggest that the learner review a particular module.

- In the design of control features in e-learning systems, provide as much pacing control as possible through features that control the pace at which learning proceeds. These features can include use of the "forward" and "back" buttons, "replay" buttons for video presentations, and "quit" and "continue" options. It is essential that learners be aware of where all functions that can affect the pace of learning are situated.

- Ensure that e-learning system features related to navigation such as topic headers, topic menus, course or site maps, and links to other resources are perceptually salient and understood. Use links with caution and sparingly.

- In e-learning systems that contain gaming features, avoid "twitch" features that require learners to provide quick and precise movements in response to a displayed condition.

- Try to match the type of game with the nature of the learning goals. For example, adventure style games are more suited to training on troubleshooting or problem solving, whereas television-style game shows that emphasize factual knowledge are more conducive to learning about facts, categorization of information, and concrete concepts.

- Build guided feedback into simulations or games through either explanatory feedback or as hints appearing between simulation trials. This feedback during the course of the game or simulation should explain the principles or concepts that are being illustrated. Avoid corrective-only feedback.

- In e-learning game-type systems, avoid an emphasis on achieving high game scores unless achieving such goals can be clearly linked to the achievement of better learning.

- Provide the learner with the opportunity to reflect on correct responses to allow for deeper learning about the instructional content. Avoid having the learner reflect on incorrect responses as this may induce cognitive overload.

- A simulation or game should always begin with a relatively simple task or goal to allow confidence to be gained and extra cognitive processing to be available for managing any complexity that might still be arising from learning some of the features of the system.
- If a simulation or game is highly complex in its functionality, consider turning some of the features off until the learner has mastered the most fundamental features (i.e., consider using a training wheels approach).
- Consider use of a pedagogical agent that demonstrates at the outset how a game is played or how to interact with the simulation.
- Records of actions taken over the course of a simulation or game or other relevant data should be easily accessible by the learner to enable reasoning about the instructional material.
- Find creative ways for demonstrating cause-and-effect, temporal, or sequential relationships in the learning material by using sequences of static snapshots that the learner can easily control.
- Transition from worked examples to full problems using scaffolding, or what is also known as a fading approach. In this approach, the initial worked example is fully worked out or demonstrated, and then the burden is gradually shifted to the learner to complete more of the example's steps until the learner completes the problem on his or her own.
- Err on the side of providing more worked examples rather than fewer, prior to transitioning the learner to practice exercises.
- Include self-explanation questions that are in close spatial proximity to the worked-out steps of the worked examples, for example, by querying the learner to choose, from multiple choices, the appropriate explanation for a given worked example solution step.
- Supplement worked examples with explanations; that is, do not assume the learner understands the solution steps that are shown. The explanations should not be too long or complex and should be positioned in close spatial proximity to the worked-out step. As the learner gains expertise, the presentation of explanations could be made optional or in response to an error that the learner makes on a step.
- Worked examples designed to support near-transfer learning should be in the form of demonstrations that contain as much of the context as possible from the situations that the training or instruction is trying to emulate. In designing worked examples for far-transfer learning, provide examples possessing different contexts or story lines that can collectively induce flexibility in thinking in order to adapt to different problem situations.
- With worked examples, avoid incorporation of the audio modality in the form of narration.

- For practice problems, administer distributed practice, both within and across practice sessions, as opposed to mass practice.
- For practice on tasks or problems that learners will perform in job situations, tailor the practice problems to the nature of the performance requirements. Use task analysis to specify the physical and cognitive requirements of the task as it will be performed in the job environment.
- In practice problems, provide explanatory feedback that is intended to allow the learner to build correct mental models, understand concepts, or guide the correct approach or procedure.
- Consider incorporating avatars into e-learning environments.

Recommended reading

Czaja, S.J., Sharit, J., Hernandez, M.A., Nair, S.N., and Loewenstein, D. (2010). Variability among older adults in Internet health information-seeking performance. *Gerontechnology.* 9: 46–55.

Fletcher, C. and Bailey, C. (2003). Assessing self-awareness: Some issues and methods. *Journal of Managerial Psychology*, 18: 395–404.

Ketcham, C. and Stelmach, G. (2004). Movement control in the older adult. In R.W. Pew and S.B. van Hemel (Eds.), *Technology for Adaptive Aging*. Washington, DC: National Academies Press, 64–92.

Michas, I.C. and Berry, D.C. (2001). Learning a procedural task: Effectiveness of multimedia presentations. *Applied Cognitive Psychology*, 14, 555–575.

Mikropoulos, T.A. (2006). Presence: A unique characteristic in educational virtual environments. *Virtual Reality*, 10: 197–206.

Morse, S., Littleton, F., Macleod, H., and Ewins, R. (2009). The theatre of performance appraisal: Role-play in Second Life. In C. Wankel and J. Kingsley (Eds.), *Higher Education in Virtual Worlds: Teaching and Learning in Second Life*. Bingley, UK: Emerald Group , 181–201.

Shah, P. and Freedman, E.G. (2003). Visuospatial cognition in electronic learning. *Journal of Educational Computing Research*, 29: 315–324.

Sharit, J. (1997). Allocation of functions. In G. Salvendy (Ed.), *Handbook of Human Factors and Ergonomics*, 2nd ed. New York: John Wiley & Sons, 301–339.

Sharit, J., Hernandez, M.A., Czaja, S.C., and Pirolli, P. (2008). Investigating the roles of knowledge and cognitive abilities in older adult information seeking on the web. *ACM Transactions on Computer-Human Interaction*, 15(1): 3–25.

Sharit, J., Hernandez, M.A., Nair, S.N., Kuhn, T., and Czaja, S.J. (2011). Health problem solving by older persons using a complex government website: Analysis and implications for web design. *ACM Transactions on Accessible Computing*, 3(11): 1–35.

Stanney, K.M. and Salvendy, G. (1995). Information visualization; assisting low spatial individuals with information access tasks through the use of visual mediators. *Ergonomics*, 38: 1184–1198.

Van Bruggen, J.M., Kirschner, P.A., and Jochems, W. (2002). External representation of argumentation in CSCL and the management of cognitive load. *Learning and Instruction*, 12: 121–138.

Zhang, J. (1997). The nature of external representations in problem solving. *Cognitive Science*, 21: 179–217.

chapter eleven

Performance assessment and program evaluation

11.1 Introduction

Chapter 9 presents an overview of a systematic design model for developing a training program. This model indicates that there should be a logical flow to program design and incorporates four basic phases: a front-end analysis phase, a design and development phase, a full-scale development phase, and a final evaluation phase. Implicit in this model is the incorporation of a user-centered design approach that involves usability testing and feedback from representative users of the training program. In many cases users include both instructors and trainees, given that instructors who actually deliver training programs may not have been involved in the initial design of the program. Think of our earlier example of the community computer/Internet training course (Chapter 3) where the course was taught by community trainers. In this case, obtaining feedback about the program from representatives of both groups of users would be important. As noted below, in this case it would also be helpful to get feedback from administrative staff at the community agencies. The user-centered design approach is also an iterative design approach where feedback obtained from users is used to modify the design of a system, in this case the training program (see Figure 9.1).

The *front-end analysis phase* includes an organizational analysis, job/task analysis, and trainee analysis. This phase of the process is used to define the requirements and constraints of the training program and is critical to the specification of the learning objectives, the format and content of the program, implementation plan, and the evaluation criterion. In essence, the analyses conducted in this phase provide information on the organizational (e.g., organizational climate, goals), environmental (e.g., classroom, workplace, home), and social (e.g., peer support available) context in which the training will occur; what needs to be trained (knowledge, skills, attitudes, behaviors); and who needs to be trained (novices, experts, younger adults, older adults). A front-end analysis is critical to the development of an effective training program. Training programs will have limited success if they are based on a "one size fits all" design approach.

The *design and development phase* uses the information obtained in the front-end analysis to develop a model or blueprint of the training program. In this phase, training goals and objectives are specified, instructional strategies and methods are selected, training content is developed, and other design decisions are made about duration of training, equipment requirements, feedback mechanisms, and practice schedules. General guidelines for older adults regarding these issues are reviewed in Chapter 3. This phase is usually an iterative process where the designers generate a number of design concepts or prototypes and conduct some formative evaluations that include usability testing and often a cost–benefit analysis. The formative evaluation may involve select aspects of the training program such as the pacing constraints of an online program. However, at some point it is advisable that feedback be obtained on the entire program and the training materials (e.g., instruction manuals, assessment materials). It is important that the training or learning objectives developed in this phase are specific, measureable, and relevant to the task. For example, if an objective of the training program is that the learner be able to perform basic tasks using a text-editing program such as Word, it is important to specify the nature of the tasks and what constitutes performance success. This might mean that the learner must be able to open a new document file, enter text into the file, perform some editing functions such as cutting and pasting text, and save the file. Furthermore, they must be able to do this without instructor assistance. We return to this issue later in the chapter. Once the training program is fully developed it is ready for full-scale implementation. This often involves activities such as training of instructors, preparing the environment (e.g., ensuring Internet access), and communicating program logistics with the trainees.

The final phase of training program design is the *evaluation phase*, which is the focus of this chapter. Evaluation is a critical aspect of training design and is imperative with respect to determination of the success of a training program. The evaluation plan should be systematic and address issues regarding what needs to be measured, when measurements need to occur, who should be assessed and by whom, and where (what context) the assessments should occur. For example, if a sample of older adults is being trained in basic technology skills for workplace re-entry and is receiving training on a text-editing and a spreadsheet application, one needs to consider if performance should be measured after each module as well as after completion of the entire course; what measures should be used to measure performance; who should administer the assessments, the trainers or someone independent of training delivery; and if the assessments should be restricted to the training environment or if it might be possible to also gather information from the community organization or a worker placement program that is assisting the trainees with employment. If possible, it would also be helpful to gather feedback about the training at a later point in time from the trainees or an organization depending on the goals of the evaluation. In our evaluation of the community-based computer/Internet training program we conducted a short telephone interview with the trainees to examine changes in their overall use of technology, participation in additional computer training programs, and their attitudes toward technology.

The data gathered in the training evaluation are used to systematically assess the need for changes to the design of the program. For example, in our evaluation of the community-based computer/Internet training program we learned that the trainees felt that the course should include more interactive practice examples and hands-on assistance. The program has since been adapted to accommodate this feedback. An important aspect of a systems approach to training design is an emphasis on a continuous assessment of needs and gathering of evaluative feedback to ensure that a training program is meeting its stated objectives. For example, a training program designed to teach skills for some software application would not be effective if the program were not adapted to include changes in the software or ignored feedback from the instructors and the trainees about the need for more practice on basic constructs. In other words, training programs must not be viewed as static but instead as dynamic and adaptable. In fact, evaluation is implicit in every phase of the design process.

Unfortunately, for a variety of reasons evaluation of training programs is often incomplete or ignored in practice. Reasons for neglecting this aspect of program design include budget and time constraints, lack of expertise, and lack of methods or tools for conducting the evaluation. In the following section we present some commonly used models that have evolved to guide the evaluation process. We then present some examples of assessment metrics. Finally, we discuss factors that need to be considered when conducting assessments with older adults. We begin with a definition of evaluation.

11.2 Models and approaches to training evaluation

11.2.1 Defining evaluation

Evaluation refers to a systematic process for measuring the outcomes of training programs or the extent to which training goals were met. This construct is somewhat complex, as evaluation can occur on a variety of levels and include an evaluation of trainee learning, instructional materials, transfer of training, organizational outcomes, and cost effectiveness. Also, the goals of training programs vary tremendously and can include outcomes such as imparting new skills or refreshing existing skills (e.g., a software application, a language), increasing knowledge about a particular topic (e.g., cultural competence), changing behaviors (e.g., exercise, safety habits), or changing attitudes or perceptions (e.g., self-efficacy). Whatever the goals of the training program, implicit in a systems approach to training design is a systematic approach to evaluation where the objectives and methods of the evaluation are carefully planned and implemented.

Another important distinction that arises when discussing evaluation is between the constructs of *training evaluation* and *training effectiveness*. Training evaluation is a measurement process that examines the extent to which a training program meets the intended goals, whereas training effectiveness is concerned with measuring the particular factors that influence training outcomes such as the environment, elements of the program, or characteristics of the trainees. In other words, training evaluation asks, "Did the program work," and training effectiveness asks, "Why" the program was successful or unsuccessful. For example, in our evaluation of the community-based computer/Internet training course we addressed the issue of whether the program worked in terms of increasing the trainees' knowledge and skills and changing their attitudes toward computers. We did not compare two versions of the program with different characteristics such as a face-to-face program versus an online program to determine which method of training was more effective. We did, however, include a wait-list no-training control group. The inclusion of this group strengthened the design of our evaluation protocol. In contrast, in our study of training for the Medicare website we examined two methods of training: a unimodal online approach and a multimedia online approach. In addition, we included a no-training control group. Thus, in that study we were able to address the general question of whether training in and of itself improved the ability of the participants to use the website. We were also able to assess whether a unimodal approach was more effective (resulted in greater performance gains) than a multimedia approach. Ultimately, assessments of both training outcomes and training effectiveness provide important insights for improving training programs.

11.2.2 *Training evaluation models*

To help with the development of methods and tools for training evaluation different models have been proposed. We provide a brief discussion of these models as they help identify the levels or aspects of training that need to be evaluated as well as assessment metrics. One of the initial models for training evaluation is Kirkpatrick's (1976) multilevel model approach that discusses evaluation in terms of four levels or criterion types: (1) trainee reactions (what the trainee thinks about the program), (2) learning (what the trainee has learned), (3) behavior (how the trainee's behavior changes), and (4) organizational results (impact on the

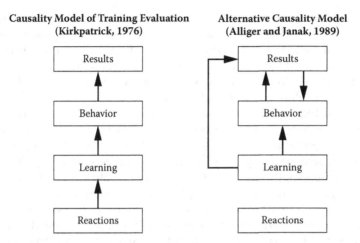

Causality Model of Training Evaluation
(Kirkpatrick, 1976)

Alternative Causality Model
(Alliger and Janak, 1989)

Figure 11.1 A comparison of two training evaluation models. (From Salas et al., in *Handbook of Human Factors and Ergonomics* 2006. Hoboken, NJ: John Wiley & Sons. With permission.)

organization or training provider). As shown in Figure 11.1, reactions are related to learning, learning is related to behavior, and behavior is related to results. This model is one of the simplest approaches for understanding training evaluation and for this reason has been widely used as a framework to guide evaluation efforts. Criticisms of the model include the simplicity of the model and the fact that other important outcomes such as trainee motivation and self-efficacy are ignored, as well as outcomes such as the validity of the trainee content and cost-effectiveness.

In response to these weaknesses several alternative models of training evaluation have evolved. One was developed by Kraiger and colleagues (Kraiger, Ford, and Salas, 1993; Kraiger, 2002) and emphasized three multidimensional target areas for evaluation: training content and design, changes in learners, and organizational. In this model, reactions are related to training content and design and measure how well the trainees liked the program as well as their perceptions of program utility. According to this model, reactions are not related to changes in learners or organizational payoffs.

Alliger and Janak (1989), on the basis of a meta-analysis of Kirkpatrick's approach, also proposed an alternative to Kirkpatrick's model. Their model also included four levels for evaluation: reactions, learning, behavior, and results. However, they delineated different types of reactions, affective and utility judgments, and divided learning into the three categories of immediate knowledge, knowledge retention, and behavior/skill demonstration, and also focused more on transfer to the work environment rather than the more general behavior category. In addition, they posited, similar to Kraiger and colleagues, that there is little or no relationship between reaction criteria and other levels of training such as learning. In other words satisfaction with a training program has very little relationship to whether a person has learned something from training. However, there is a positive relationship between learning and behavior and between behavior and results (Figure 11.1).

Clearly, the models discussed are highly similar and underscore the complexity of the evaluation process and the fact the evaluation needs to be multidimensional and target different aspects of the training program. It is insufficient to simply measure learning or reactions. At the very least, the evaluation should include metrics of both outcomes as the evidence suggests positive reactions among trainees do not guarantee that learning has taken place or that changes in behavior will be exhibited outside the training environment. However, this is not to suggest that trainee reactions and feedback are not important. As discussed throughout this chapter, they are a critical component of training evaluation and provide important information for redesign of training programs.

Based on our earlier discussion of the important distinction between training evaluation and training effectiveness we present a final model of training evaluation that combines evaluation and effectiveness. The model was developed by Alvarez and colleagues (Alvarez, Salas, and Garofano, 2004) on the basis of an extensive review and integration of the literature on training evaluation and effectiveness. Although the model may appear to be complex, we include it in our discussion to help elucidate the differences between the constructs of evaluation and effectiveness and to provide a framework for assessment of training outcomes. This model also helps to demonstrate how the various elements of a training program and training outcomes interact and influence one another.

The model is consistent with a systems design approach, and as shown in Figure 11.2, incorporates four levels. The top level is the needs analysis, which is included because it is used to develop training objectives, design, and content, and is related to the targets of training evaluation such as changes in learners and organizational results. The second and third levels combine recent models of training evaluation and measures that are commonly used in training research. According to the model, training content and design can be evaluated by measuring reactions to training; changes in learning can be assessed by measuring post-training attitudes, cognitive learning, and training performance; and organizational outcomes can be determined by measuring transfer performance (e.g., behavioral changes on the job) and results (e.g., increased job safety performance, morale, work efficiency, continued use of technology).

The fourth level of the model is related to training effectiveness and indicates that training outcomes cannot be evaluated without

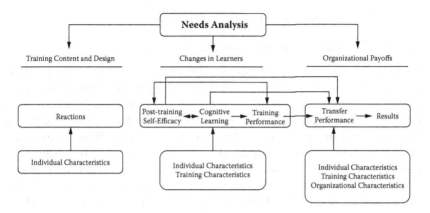

Figure 11.2 An integrated model of training evaluation and effectiveness. (From Alvarez, Salas, and Garofano, 2004. *Human Resource Development Review,* 3: 385–416. With permission.)

consideration of individual trainee characteristics, training design elements, and the organizational or situational context. For example, older adults who have experience using computers and the Internet may rate an Internet training course that allocates a large percentage of time to basic computer skills less positively and find the training to be of limited utility to their learning goals as compared to individuals with less experience. Post-training attitudes, cognitive learning, and training performance are also influenced by individual characteristics as well as the characteristics of the training program. For example, self-efficacy (one's belief in one's own competence; ability to achieve certain goals), such as computer self-efficacy, is a common measure of post-training attitudes and is influenced by factors such as feedback provided during training. Cognitive learning and training performance are influenced by factors such as feedback, practice schedules, learner involvement, and the like. Also, a program that is successful in a classroom setting where learners have peer support may be less successful in a home environment where peer support is lacking. Transfer performance is related to factors such as the variety of examples provided in training, amount of practice, degree of fidelity between the training and transfer environments, and the degree to which the post-training environment is supportive of the application of new skills. Trained and learned skills will not be demonstrated as performance outcomes (e.g., job-related behaviors) if trainees do not have an opportunity to perform them. Overall, the model developed by Alvarez and colleagues provides a comprehensive framework for evaluating training programs and illustrates that training outcomes cannot be viewed in isolation of contextual factors.

11.3   Assessment of training outcomes

11.3.1   Criteria for selection of outcome measures

In the prior section we discussed models of training evaluation. Although the models varied somewhat with respect to emphasis, complexity, and hypothesized relationships among the model components, all of the models clearly underscore the fact that the evaluation of a training program should be systematic and target its different aspects. Of course, there may be situational constraints that limit the extensiveness of an evaluation plan. However, as suggested earlier, it is generally not sufficient simply to measure trainee reactions or post-training performance alone. The training evaluation plan should include a specification of who is conducting the assessments, the protocol and time schedule for the assessments, and the measures that will be used in the evaluation. For example, decisions need to be made regarding the frequency of assessments, such as if they will be restricted to pre- and post-training or include a measure of transfer in the situational environment such as the job. Decisions also need to be made regarding how post-training performance will be assessed, for example, will it include a measure of knowledge as well as a performance-based measure and a measure in the field. It is also possible to conduct technology-based performance assessments. For example, online training programs may include performance assessments throughout the program to help tailor or adapt the program to the needs of the training. Other decisions relate to the evaluation design strategy (e.g., pre–post; post-testing only; use of a control group), sample size, and costs associated with the evaluation.

The choice of evaluation criteria is a primary decision that must be made during the planning of the evaluation phase. These criteria are used to judge the success of a training program and clearly statements about success can vary according to the criteria chosen to measure it. Characteristics of the data collected (e.g., type of scale used) also influence the type of methods that can be used to analyze the data, which influences the type of conclusions that can be drawn (e.g., associations between variables or cause and effect statements). Other characteristics of the data such as objectivity also affect the conclusions that are drawn from the data. Supervisory ratings of post-training job performance of older workers may be negatively biased, for example, if a supervisor maintains the belief that older workers are unable to learn new skills.

The choice of measures depends on the target of the evaluation, the evaluation objectives, and the available resources. For example, assume one were interested in determining if the pacing of a training program was stressful for older learners and decided to use physiological indices of stress to measure this outcome. Although the data might prove to be interesting, this approach would not be feasible without personnel available who are trained in the use of these assessment protocols and the technical resources needed to analyze the data. Also, the measures chosen should meet the basic measurement criteria of validity (measures what is intended to be measured), reliability (the measure is stable or consistent), and sensitivity (the ability of the measure to detect change). Other considerations are the level of objectivity, specificity (e.g., whether a measure refers to a particular task or is generic), intrusiveness, and acceptability of the measure. A detailed discussion of the statistical properties of measures and data analysis techniques is beyond the scope of this chapter. Recommendations for further reading on these topics are provided at the end of the chapter.

11.3.2   Examples of outcome measures

Table 11.1 presents some suggestions for types of metrics that might be used in an evaluation program. The suggestions are grouped according to Kirkpatrick's four levels of evaluation and include examples of what is being measured, types of measures, and sample questions to address the issue. The list is far from exhaustive. Reactive criteria are measured by self-report criteria and reflect affective and attitudinal responses to the training program. These types of measures are usually administered to the trainee and in many cases to the program instructors. Specific topics addressed by these measures include: satisfaction with the training, perceptions of the usefulness of the training, perceptions of the quality of the training, ratings of the instructor, course content and structure, course materials, and in some cases the training environment. The instructor's measures typically also include ratings of the trainee characteristics. These types of measures are given post-training and are relatively easy to administer.

Learning criteria reflect the learning outcomes of training and may also include post-attitudinal outcomes such as changes in self-efficacy or comfort using technology. This may take the form of a paper and pencil pre–post test assessment of knowledge or some attitudinal measure. It may also involve some performance-based measure using simulation or techniques such as role-playing. Behavior criteria relate to performance outside the training setting such as on-the-job performance. These types of measures include supervisory ratings of performance, observation of performance, or objective measures of performance such as greater adherence to safety standards, increased productivity, or changes in use of resources.

Behavior criteria may also include self-ratings from the training at some later point in time following training. For example, one might query someone who has completed a training course on how to use

Table 11.1 Assessment Metrics Associated with Levels of Training Evaluation Based on Kirkpatrick's Multilevel Training Typology

Level	What Is Being Measured/ Evaluated	Measurement	Sample Questions
1. Reactions	Learner and instructor reactions after training Satisfaction with training Ratings of course materials Effectiveness of content delivery	Self-report survey Evaluation or critique	Do you like the training? Do you think the trainer was helpful? How helpful were the training objectives?
2. Learning	Attainment of trained competencies (i.e., knowledge, skills, and attitudes) Mastery of learning objectives	Final examination Performance exercises Knowledge pretests and posttests	True or False: Large training departments are essential for effective training. Supervisors are closer to employees than is upper management.
3. Behavior	Applications of learning competencies on the job Transfer of training Improvement in individual or team performance	Observation of job performance	Do the trainers perform learned behaviors? Are the trainers paying attention and being observant? Have the trainees shown patience?
4. Results	Operational outcomes Return on training investment Benefit to organization	Longitudinal data Cost–benefit analysis Organizational outcomes	Have there been observable changes in employee turnover, employee attitudes, and safety since training?

Source: Adapted from Salas et al., in G. Salvendy (Ed.) *Handbook of Human Factors and Ergonomics*, 3rd ed. Hoboken, NJ: John Wiley & Sons, 2006.

Internet health websites if they have increased their use of these websites to gather health information. It might also involve comparing the behavior of people who have received the training versus those who have not received the training. Sometimes obtaining objective measures of performance or observing performance in the field can be challenging. For example, it may be difficult to have access to objective measures of work output or efficiency. Group field comparisons also present special challenges with respect to factors such as sample equivalence. In addition, caution needs to be exercised when relying on supervisory or self-ratings of performance, as these types of ratings are subjective and may be prone to bias. Finally, results criteria refer to the impact of the training on an organization or agency sponsoring the training. These types of criteria might include measures of productivity, cost–benefit ratios, improved safety, and other benefits to an organization such as changes in quality ratings or customer satisfaction. The results of these assessments are often difficult to interpret as they are the most distal from training and may be influenced by a wide variety of other factors.

In most cases evaluation should include a variety of metrics and subjective as well as objective measures. In the selection of measures it is also a good idea to begin with a search of the relevant literature to identify frequently used measures. This is helpful as it avoids the need to re-invent the wheel and also provides results that can be compared with other evaluation studies. Most of these measures also have known psychometric properties. Sometimes there is a need to develop a new measure or to adapt an existing measure. In our study of multimedia training we included a standardized measure of computer attitudes pre- and post-training and also developed a measure to evaluate participant's reaction to the training specific to the program. Of course some of the questions used in this measure were adapted from other similar measures. In all cases pilot testing is a critical aspect of measure selection to ensure that you are measuring what you are intending to measure; the measurement protocols are feasible; and the measures are understandable and acceptable to representative samples of individuals who will be included in the evaluation. As discussed above, a comprehensive evaluation also requires the consideration of the influence of contextual variables on training outcomes.

11.3.3 Including older adults in assessment protocols

We conclude this chapter by discussing some unique characteristics of older populations that should be considered when involving older adults in evaluation. As noted throughout this book, older adults are very heterogeneous and thus it is important that representative samples of older adults on characteristics relevant to the training are included in evaluation assessments and pilot testing of evaluation instruments. For example, assume one decided to collect performance data from a sample of older adult trainees who completed a basic work re-entry training program and the trainees ranged in age from 50–75 years. In this case it would be important to include individuals in their 50s and 70s as the abilities and experiences of those in their 50s are likely to be different from those in their 70s. Health and ability must also be considered. For example, in some studies we have conducted post-training telephone interviews and found this to be challenging for some older adults with hearing deficits who have difficulty understanding telephone conversations, especially if the language is technical in nature or they are not used to phone-based assessments. In these cases we have conducted the interviews in person or used a paper-and-pencil version administered via mail.

In general, the design and formatting of assessment instruments is an extremely important consideration for older adults. Care must be taken to ensure existing guidelines for formatting text and speech information and factors such as pacing are followed. It is also important that highly technical language is avoided and that familiar vocabulary is used so that the content of the assessment is understandable to people with varying backgrounds. If possible, the use of instruments that have been *normed* with older adults (the measure has been standardized with representative samples of older adults) is recommended. In any case all measurement instruments should be pilot tested prior to formal data collection. Protocols for assessment should also be pilot tested.

Stamina also decreases with age, and thus extra consideration needs to be given to participant burden and fatigue. Lengthy testing sessions should be avoided and frequent rest breaks should be provided. Older adults are also less likely than younger adults to have recent experiences in a training/assessment environment and thus they may have some anxiety about their performance and being evaluated. Thus it is important to make the evaluation environment as stress-free as possible and, given age-related changes in processing speed, extra time should be allowed for responding. In addition, it is important to clarify that the assessment is being conducted to evaluate the training program and identify areas where the program may need to be improved or changed. Finally, evaluators or individuals conducting the assessments should receive some basic training about aging and older adults. This may help eliminate biases or stereotypes that may influence administration of the assessment instruments.

11.4 Summary

The development of training programs should be based on a user-centered systems design approach. Evaluation is a critical phase of the training process and provides information about the success of a program in terms of meeting stated objectives and goals and provides information about needed changes to the design of the program. Evaluation should be an ongoing process as the characteristics of trainees, training content, and learning and performance context are dynamic and likely to change. An evaluation plan should be systematic and address issues regarding what needs to be measured, when measurements need to occur, who should be assessed and by whom, and where (what context) the assessments should occur. Evaluation should also be multidimensional and reflect different aspects of the training program. In this chapter we provided an overview of some the issues that need to be considered when developing an evaluation plan. We also presented some examples of tools and methods that can be used in evaluation. A summary of the information discussed in this chapter is presented below.

- There should be a logical flow to training program design that incorporates four basic phases: a front-end analysis phase, a design and development phase, a full-scale development phase, and a final evaluation phase.
- The design of training programs should be a user-centered design approach that involves usability testing and feedback from representative users of the training program.
- Training programs should be viewed as dynamic, evolving, and adaptive.
- Evaluation is a critical aspect of training design and it is imperative with respect to determination of the success of a training program.
- A training evaluation plan should be systematic and address issues regarding what needs to be measured, when measurements need to occur, who should be assessed and by whom, and where (what context) the assessments should occur.
- Implicit in a systems approach to training design is a systematic approach to evaluation where the objectives and methods of the evaluation are carefully planned and implemented.
- An important distinction is between the constructs of training evaluation and training effectiveness. Training evaluation measures the extent to which a training program meets the intended goals, whereas training effectiveness assesses the factors that influence training outcomes.
- An evaluation plan needs to be multidimensional and target different aspects of the training program.
- Training evaluation plans should include a specification of who is conducting the assessments, the protocol and time schedule for the assessments, and the measures that will be used in the evaluation.
- The choice of evaluation measures should be based on the target of the evaluation, the evaluation objectives, and the available resources.
- When selecting measures, begin with a search of the relevant literature to identify standardized measures with known psychometric properties that are frequently used in assessment.
- The design and formatting of assessment instruments is an extremely important consideration for older adults; adhere to existing guidelines for formatting text and speech information.

- As far as possible use nontechnical and familiar language in measurement instruments.
- Measurement instruments and protocols for assessment should also be pilot tested with representative samples of older adults prior to formal data collection.
- Lengthy testing sessions should be avoided and frequent rest breaks should be provided.
- The evaluation environment should be as stress-free as possible and individuals should be allowed sufficient time to respond.

Recommended reading

Edwards, P.J., Sainfort, F., Kongnakorn, T., and Jacko, J.A. (2006). Methods of evaluating outcomes. In G. Salvendy (Ed.), *Handbook of Human Factors and Ergonomics*, 3rd ed. Hoboken, NJ: John Wiley & Sons, 1150–1187.

Fisk, A.D., Rogers, W.A., Charness, N., Czaja, S.J., and Sharit, J. (2009). *Designing for Older Adults: Principles and Creative Human Factors Approaches*, 2nd ed. Boca Raton, FL: CRC Press.

Kirkpatrick, D.L. and Kirkpatrick, J.D. (1994). *Evaluating Training Programs*. San Francisco: Berrett-Koehler.

Salas, E., Wilson, K.A., Priest, H.A., and Guthrie, J.W. (2006). Design, delivery, and evaluation of training systems. In G. Salvendy (Ed.), *Handbook of Human Factors and Ergonomics*, 3rd ed. Hoboken, NJ: John Wiley & Sons, 472–512.

Vercruyssen, M. and Hendrick, H.W. (2011). *Behavioral Research and Analysis: An Introduction to Statistics Within the Context of Experimental Design*, 4th ed. Boca Raton, FL: CRC Press.

chapter twelve

Conclusions and synthesis

12.1 Introduction

The goals of this book were to present a state-of-the-science summary of the topic of training and instructional design for older adults and to present some basic principles and guidelines regarding "best practices" for designing training and instructional programs for older people. As demonstrated by the wealth of material presented in the previous chapters, the topic of training and instructional design is very broad, and thus this book covers a wide range of issues. The information provided represents a compilation of current knowledge and research about aging, learning and skill acquisition, training and instructional design, and our experience. Our approach was to summarize current thinking regarding training and instructional design and to demonstrate, based on existing knowledge about aging, how these findings can be applied to guide the development of training programs for older people. Of course we also based our recommendations on the rather limited amount of research that specifically addresses issues related to training and older adults. In some cases we were quite specific; in other instances the recommendations were by necessity more general in nature. Recommended readings were also given to provide more depth in the various topics. Our orientation was based on a human factors approach, which maintains that training and instructional programs must be designed with consideration of the training population, the tasks to be trained, and the context in which the training will occur.

The focus of this chapter is on the general themes that emerged throughout this book. Before we begin our synthesis of themes, to emphasize the importance of the topic of aging and training we provide some additional discussion of various domains where training is likely to be important for older adults. As becomes evident throughout this discussion, training is and will continue to be a critical component of most domains of everyday living.

12.2   Exemplar applications

12.2.1   Work and employment

As noted in Chapter 2, for a variety of reasons many older adults are choosing to remain in the workforce longer, re-entering the workforce, or beginning a second career. Given the dynamic nature of the workplace and the infusion of technology into most jobs and work environments, workers of all ages, but especially older workers, will need to continually upgrade their skills and knowledge to remain competitive in the workplace. This will involve training to learn new tools and equipment, new ways of doing tasks, new software applications, or a new job. It may also involve learning new ways of communicating with coworkers using tools such as e-mail, instant messaging, or videoconferencing.

For example, routine production functions have become streamlined and increasingly automated, which has important consequences for the skill requirements of manufacturing workers. Traditionally, the emphasis in these tasks was on physical demands, whereas today the emphasis has become more on cognitive demands and technology skills. Healthcare workers also increasingly have to interact with a variety of types of medical devices and equipment; technology has become ubiquitous within the healthcare domain. The transition to electronic medical records (EMRs) implies that doctors, nurses, and other type of providers will need to learn different means of creating, maintaining, and tracking patient records. In addition, e-mail is becoming a more common form of communication between providers and consumers. The Internet is also becoming a common mechanism for providers to gather information on an illness or disease. Customer service workers traditionally interacted with customer queries over the telephone and often relied on hardcopy records to respond to queries. This type of task now often involves the use of electronic databases and e-mail. Service sector workers such as those who work in restaurants, fast-food chains, or in sales are also increasingly required to conduct customer transactions using technology. The examples of technology use by workers and the resultant changes in job demands and skill requirements are endless and will continue with developments in technology.

Overall, the nature of work has changed and will continue to change creating high demand for worker training and retraining. In addition, the nature of training is changing and organizations are increasingly relying on e-learning formats such as online learning programs and virtual classrooms to meet their training needs. Thus, workers need even basic technical skills to interact with these training formats to receive the work-related training that they require. This can be challenging for many workers, especially older workers who are more likely to have less familiarity with these types of learning formats.

12.2.2   Healthcare and living environments

As noted, technology is pervasive within the healthcare arena for both providers and healthcare consumers. As healthcare moves outside clinical settings and into the home, consumers are increasingly confronted with the need to learn to use complex technologies and devices. For example, people with chronic disease need to interact with devices such as lifelines, blood glucose monitors, blood pressure monitors, infusion pumps, or ventilators. They may also be linked to a telemedicine system so that their providers can monitor their health status or make virtual house calls. These types of systems and devices are typically used without the supervision of a professional. Imagine, for example, an older person with diabetes and hypertension who lives alone and needs to use a blood pressure monitor and blood glucose meter that is linked to a health monitoring system. These types of systems can be complex in terms of operating procedures and require calibration, coordination among devices, and maintenance. Thus training and well-designed instructional materials are essential to their use.

The Internet is also becoming a prevalent platform for the delivery of health information and services. There are a vast and steadily growing number of health websites, and access to many health services and resources now requires use of the Internet. In the near future consumers from all types of demographic groups will also be interacting with patient portals tethered to an EMR to perform tasks such as reordering prescriptions, making appointments, communicating with their providers, and monitoring their health status. Many of these applications are also being tethered to personal communication devices such as smart phones. Technology is also being used to deliver intervention programs to patients and family caregivers. For example, our team has been involved in the development and implementation of a technology-based caregiver intervention program for family caregivers of dementia patients that involves the use of videophones. We were also involved in the implementation of an intervention program for troubled adolescents and their families that was delivered via netbooks. Health maintenance programs such as stress management, and physical and cognitive exercise programs are also frequently presented online. Use of these applications requires training in the use of the technology as well as training in other aspects such as credible sources of health information and basic health literacy constructs.

Home environments are also being equipped with other types of technology systems such as home monitoring and security systems, complex environmental control and entertainment systems, and food preparation devices (e.g., microwaves, coffee makers). Furthermore, many times these systems are integrated through some type of master control device. It is not uncommon for a lifeline device such as a panic button to be connected to a home alarm system. Also, oftentimes user manuals for these products and devices are now online. Clearly, use of these systems, products, and devices requires training and the availability of good instructional materials.

12.2.3   Transportation, communication, and other everyday domains

Everyday transportation is also becoming more complex and requiring instruction even for frequent travelers. Automobiles have many more features for controlling factors such as seat comfort, temperature, driving speed, and entertainment. As such, automobile displays and controls are becoming more complex. Automobiles are also incorporating advanced technology systems such as directional systems (GPS), distraction management systems, and collision avoidance warning systems. Operating and driving an automobile now requires learning the rules of the road as well as learning how to use the functionality embedded in cars and advanced intelligent system technologies. In fact, some automobiles now come with a user manual as well as an accompanying DVD.

Use of other forms of transportation such as airplanes, buses, and trains are also placing new demands on consumers. Tasks such as obtaining travel schedules, making reservations, and printing boarding passes typically involve the use of some type of automated system such as a kiosk and fewer interactions with an agent or customer service representative. Unfortunately, there is often variation in how these systems operate across providers or carriers. There are also varying protocols with respect to security issues. Thus consumers need basic orientation or some form of instruction regarding the various aspects of travel logistics.

Money management and banking tasks are also becoming more technology-oriented and involve fewer face-to-face interactions. The use of automatic teller machines (ATMs) is ubiquitous and ATMs are being used for a wider variety of banking applications beyond simple cash disbursement. Thus, for today's cohort of older adults this represents new ways of performing familiar tasks. Data from our group suggest that older adults may be more willing to use these types of systems if they had some form of training. Paying bills online and online shopping are also becoming more common. In fact, some retailers only have online stores.

Communication is also changing. Cell phones and smart phones that combine mobile phone functionality and elements of computing, and other types of devices such as iPads, are becoming commonplace. These devices support activities well beyond communication activities and include activities such as searching the Internet, healthcare interventions (e.g., medication reminders or nutrition/exercise tracking), checking flight status, money management, games, reading books or newspapers, and watching

TV shows or movies. The types of communication media supported by these devices include a variety of formats such as texting, tweeting, and e-mail. Although the types of technologies discussed in this section may be less complex and cognitively demanding than those discussed in the sections on work or healthcare, use of these devices still requires new learning on the part of the user, especially because these technologies are constantly being upgraded to provide more functionality. In addition, training for many of the applications discussed requires training on how to use the application and training on other issues such as privacy and security as well as new rules for etiquette (e.g., e-mail etiquette).

12.3   Themes

12.3.1   Older adults represent an important training population

People are living longer and remaining more active as they age. By 2030 there will be about 1 billion people aged 65+ (Administration on Aging, 2007) and increased numbers of people who are living into their 80s, 90s, and even 100s. As the population ages there is a parallel increase in the population of older people who need or desire to engage in new learning or brush up on previously learned skills. As illustrated in the previous section, there are a wide variety of circumstances where older adults will need some form of training to keep pace with developments in technology, use new products or devices, and keep abreast of changes in the workplace and other domains such as the service sector and healthcare. A majority of older people are also actively interested in pursuing training and learning opportunities in order to remain active and engaged and pursue personal interests. Therefore, the needs, characteristics, experiences, and preferences of older people must be considered in the design of training programs and support materials.

12.3.2   Older adults represent a unique and heterogeneous population

The process of aging is accompanied by changes in sensory/perceptual processes, cognition, motor skills, and functional capacity. Some of these changes such as changes in vision and auditory acuity, stamina, attentional capacity, processing speed, and working memory represent declines. However, there are also abilities that show little decline or even increase with age such as vocabulary, experience, and wisdom. Aging may also be accompanied by social and economic changes such as in living arrangements or income.

As stressed throughout this book, older adults are very heterogeneous and vary tremendously with respect to demographic characteristics, life experiences, health status, skills, and abilities. There are vast differences between and within cohorts of older adults. Aging is also associated with tremendous plasticity and older people can experience improvements and gains in physical, cognitive, and functional performance and can learn new skills. Also, the fact that older adults bring a wealth of knowledge, skills, and experiences to situations should not be overlooked. Designers of training programs need to ensure that such programs accommodate the diversity of the older adult population and that pilot testing and evaluation of these programs is conducted with representative older adult user groups. In addition, where possible, training programs should build on the prior experience and expertise of older adults.

12.3.3   Older adults are willing and able to learn

Another important theme of this book is that older adults are capable of and willing to learn new skills, technologies, and activities. They are also interested in engaging in learning and training activities. However, just as with learners of all ages, they must be motivated to engage in the learning process, feel comfortable in the training situation, and receive the proper type of instruction.

Generally, we know that older adults typically take longer to learn new skills, and require more practice and environmental support. In addition, they may fatigue more easily and experience more anxiety than younger people in learning situations. These factors need to be accounted for in the design of training programs. However, the take-home message is that older people, even those in the later decades, can learn new skills or demonstrate gains or improvements in existing skills. Older adults must be considered as a viable target population and provided with opportunities to participate in training programs and be made aware of these opportunities. Unfortunately, in many circumstances such as the workplace older people are bypassed with respect to training opportunities because of existing myths regarding aging and learning.

12.3.4   Learning and skill acquisition is a complex process

Learning is a complex process, which involves stages where the learner engages in different types and levels of processing of the to-be-learned material. An important goal in any instructional process is to enable trainees to retain and use the knowledge and skills that they learn beyond the training period. As discussed in several chapters, the degree to which learning and the subsequent retention and transfer of knowledge occurs is influenced by a wide variety of factors that include demographic factors such as age, culture, and education; ability factors such as memory and attentional abilities; motivation; social factors such as availability of support; the structure and format of the training/instructional program; and environmental factors such as the location of training programs or technology capabilities within a home. Learning is also influenced by attitudinal factors such as self-efficacy, the belief that one can learn or accomplish the task that one is being trained to perform. Anxiety about ability to learn new skills and concepts, as well as fatigue, also have a pronounced impact on the learning process as they influence capacity for learning. The effort and costs associated with learning and participating in training programs are also important considerations. It is important to understand the myriad factors that influence learning in order to best accommodate these factors in the design of training programs.

12.3.5   The design of training and instructional programs should be based on a systems approach

The design of training and instructional programs should be based on a systems approach and incorporate four basic phases: a front-end analysis phase, a design and development phase, a full-scale development phase, and a final evaluation phase. It should also represent a dynamic iterative process and involve usability testing with representative users of the training program. Users should be part of the design process early and often.

Evaluation is a critical part of this process and is imperative with respect to determination of the success of a training program. The evaluation plan should be systematic and address issues regarding what needs to be measured, when measurements need to occur, who should be assessed and by whom, and where (what context) the assessments should occur. Evaluation should also be an ongoing process as the characteristics of trainees, training content, and learning and performance context are dynamic and likely to change over time and across situations. The selection and design of evaluation instruments and evaluation protocols are extremely important considerations for older adults.

12.3.6   Guidelines exist for designing training protocols to accommodate older adult learners

Designing training and instructional programs to promote meaningful learning has been a long-standing challenge, especially for older adults. There is not one best method or training approach; it depends on the characteristics of the training content, the trainee population, and the training environment. However, there are guidelines that can be used to help ensure that training materials and protocols accommodate the characteristics of older learners. We have attempted to summarize these guidelines throughout this book. These guidelines can be used as a starting point for the design of training programs. However, as we have stressed, pilot testing with representative user groups is a cornerstone of good design. Also, what constitutes good design for older adults is usually good design for most user groups and many times usability testing with older adults can help identify usability problems that would arise for other user groups.

12.4 Conclusions

This book is not intended to provide a prescription for how to design training and instructional programs for all older adults in all learning situations. Our overarching objectives were to illustrate the importance of this topic for current and future generations of older people and to highlight some of the issues that need to be considered by designers of training programs and researchers in this area. Clearly there is much more work that needs to be done in this domain, especially with the rapid emergence of new instructional technologies.

New perspectives on training and instruction will continue to evolve. At the same time, newer types of work and technology will emerge that may require a wider array of training considerations. In addition, there will continue to be a proliferation of new concepts in cognitive psychology and cognitive neuroscience that may reshape the boundaries defining our human information-processing capabilities. With the development of new measurement techniques such as genetics and epigenetics our understanding of the aging process will also continue to evolve. How all of these factors will come together to define the requirements of training and instructional programs remains unclear. One certainty is that older adults will continue to be an important part of the training population and successful design of training programs depends on a match among the structure and content of the program and the characteristics of the learner, the task, and the environment.

References

AARP (2000). AARP Survey on Lifelong Learning. http//assets.aarp.org/rgcenter/general/lifelong.pdf

AARP (2010). Approaching 65: A survey of baby boomers turning 65 years old. http://assets.aarp.org/rgcenter/general/approaching-65.pdf

Administration on Aging. (2007). A profile of older Americans: 2007. Washington, DC: US Department of Health and Human Services. Available online: http://www.aoa.gov/AoAroot/Aging_Statistics/Profile/2007/docs/2007profile.pdf

Administration on Aging. (2010). A profile of older Americans: 2010. Washington, DC: US Department of Health and Human Services. Available online: http://www.aoa.gov/AoAroot/Aging_Statistics/Profile/2010/docs/2010profile.pdf

Administration on Aging (2011). A profile of older Americans: 2011. http://www.aoa.gov/aoaroot/aging_statistics/Profile/2011/docs/2011profile.pdf

Allen, J., Ferguson, G., Blaylock, N., Byron, D., Chambers, N., Dzikovska, M., Galescu, L., and Swift, M. (2006). Chester: Towards a personal medication advisor. *Journal of Biomedical Informatics*, 39: 500–513.

Alliger, G.M., and Janak, E.A. (1989). Kirkpatrick's levels of training criteria: Thirty years later. *Personnel Psychology*, 42: 331–342.

Alvarez, K., Salas, E., and Garofano, C.M. (2004). An integrated model of training evaluation and effectiveness. *Human Resource Development Review*, 3: 385–416.

Andrade, A.D., Cifuentes, P., Mintzer, M.J., Roos, B.A., Ramankumar, A., and Ruiz, J.G. (in press). Simulating geriatric home safety assessments in a three-dimensional virtual world. *Gerontology & Geriatrics Education*. Available online: http://www.tandfonline.com/doi/full/10.1080/02701960.2011.611553

Atkinson, R.K., Mayer, R.E., and Merril, M.M. (2005). Fostering social agency in multimedia learning: Examining the impact of an animated agent's voice. *Contemporary Educational Psychology*, 30: 117–139.

Ausbel, D.P. (1969). A cognitive theory of school learning. *Psychology in the Schools*, 6: 331–335.

Baddeley, A. (1986). *Working Memory*. New York: Oxford University Press.

Branson, R.K. (1978). The interservice procedures for instructional systems development. *Educational Technology*, 18: 11–14.

Briggs, L.J. (1970). *Handbook of Procedures for the Design of Instruction*. Pittsburgh: American Institutes for Research.

Brooke, L. (2003). Human resource costs and benefits of maintaining a mature-age workforce. *International Journal of Manpower*, 24: 260–283.

Brown, J.S., Collins, A., and Duguid, P. (1989). Situated cognition and the culture of learning. *Educational Researcher*, 18: 32–42.

Bruner, J.S. (1966). *Toward a Theory of Instruction*. Cambridge, MA: Harvard University Press.

Callahan, J.S., Kiker, D.S., and Cross, T. (2003). Does method matter? A meta-analysis of the effects of training method on older learner training performance. *Journal of Management*, 29: 663–680.

Charness, N. (1987). Component processes in bridge bidding and novel problem-solving tasks. *Canadian Journal of Psychology*, 41: 223–243.

Charness, N. (2009). Skill acquisition in older adults: Psychological mechanisms. In S.J. Czaja and J. Sharit (Eds.), *Aging and Work: Issues and Implications in a Changing Landscape*. Baltimore, MD: Johns Hopkins University Press, 232–258.

Charness, N., Fox, M.C., and Mitchum, A.L. (2011). Life-span cognition and information technology. In K.L. Fingerman, C.A. Berg, J. Smith, and T.C. Antonucci (Eds.). Handbook of Life-Span Development. New York: Springer, 331–361.

Charness, N., Kelley, C. L., Bosman, E. A., and Mottram, M. (2001). Word processing training and retraining: Effects of adult age, experience, and interface. *Psychology and Aging* 16: 110–127.

Clark, A. (2005). *Learning by Doing: A Comprehensive Guide to Simulations, Computer Games, and Pedagogy in e-Learning and Other Educational Experiences*. San Francisco: John Wiley & Sons.

Clark, R.C. and Mayer, R.E. (2008). *E-Learning and the Science of Instruction: Proven Guidelines for Consumers and Designers of Multimedia Learning*, 2nd ed. San Francisco: John Wiley & Sons.

CTGV (1990). Anchored instruction and its relationship to situated cognition. *Educational Researcher*, 19 (6): 2–10.

Czaja, S.J. and Sharit, J. (2009). *Aging and Work: Issues and Implications in a Changing Landscape*. Baltimore, MD: Johns Hopkins University Press, 259–278.

Czaja, S.J., Charness, N., Fisk, A.D., Hertzog, C., Nair, S.N., Rogers, W.A., and Sharit, J. (2006). Factors predicting the use of technology: Findings from the Center for Research and Education on Aging and Technology Enhancement (CREATE). *Psychology and Aging*, 21, 333–352.

Czaja, S.J., Hammond, K., Blascovich, J.J., and Sweden, H. (1989). Age related differences in learning to use a text-editing system. *Behaviour & Information Technology*, 8: 309–319.

Czaja, S.J., Sharit, J., Hernandez, M.A., Nair, S.N., and Loewenstein, D. (2010). Variability among older adults in Internet health information-seeking performance. *Gerontechnology*. 9: 46–55.

Czaja, S.J., Sharit, J., Lee, C.C., Nair, S.N., Hernández, M.H., Arana, N., and Fu, S.H. (2012). Factors influencing use of an e-health website in a community sample of older adults. *Jounal of the American Medical Informatics Association.*

Czaja, S.J., Sharit, J., Ownby, R., Roth, D.L., and Nair, S.N. (2001). Examining age differences in performance of a complex information search and retrieval task. *Psychology and Aging*, 16: 564–579.

Dijkstra, K., Charness, N., Yordon, R., and Hamrick, J. (2012). The role of coping, relaxation, and age on task performance with new technology. *Technical Report*, Department of Psychology, Florida State University.

Domagk, S., Schwartz, R.N., and Plass, J.L. (2010). Interactivity in multimedia learning: An integrated model. *Computers in Human Behavior*, 26: 1024–1033.

Eisdorfer, C., Nowlin, J., and Wilkie, F. (1970). Improvement of learning in the aged by modification of autonomic nervous system activity. *Science*, 170: 1327–1329.

Federal Interagency Forum on Aging-Related Statistics (2010). *Older Americans 2010: Key Indicators of Well-Being*. Washington, DC: US Government Printing Office.

Fisk, A.D., Rogers, W.A., Charness, N., Czaja, S.J., and Sharit, J. (2009). *Designing for Older Adults: Principles and Creative Human Factors Approaches*, 2nd ed. Boca Raton, FL: CRC Press.

Fleishman, E.A. and Hempel, W.E., Jr. (1955). The relation between abilities and improvement with practice in a visual discrimination reaction task. *Journal of Experimental Psychology*, 49: 301–312.

Fleishman, E.A. and Quaintance, M.K. (1984). *Taxonomies of Human Performance: The Description of Human Tasks*. Orlando, FL: Academic Press.

Gagné, E.D. (1978). Long-term retention of information following learning from prose. *Review of Educational Research*, 48: 629–665.

Glenberg, A.M.,Wilkinson, A.C., and Epstein, W. (1992). The illusion of knowing: Failure in the self-assessment of comprehension. In T.O. Nelson (Ed.), *Metacognition: Core readings*. Boston: Allyn & Bacon.

Gyselinck, V., Ehrlich, M.-F., Cornoldi, C., de Beni, R., and Dubois, V. (2000). Visuospatial working memory in learning from multimedia system. *Journal of Computer Assisted Learning*, 16: 166–176.

Gyselinck, V., Cornoldi, C., Dubois, V., De Beni, R., and Ehrlich, M.F. (2002). Visuospatial memory and phonological loop in learning from multimedia. *Applied Cognitive Psychology*, 16: 665–685.

Hambrick, D.Z., Salthouse, T.A., and Meinz, E.J. (1999). Predictors of crossword puzzle proficiency and moderators of age-cognition relations. *Journal of Experimental Psychology: General*, 128: 131–164.

Human Factors and Ergonomics Society (2012). http://www.hfes.org

Jastrzembski, T.S. and Charness, N. (2007). The model human processor and the older adult: Parameter estimation and validation within a mobile phone task. *Journal of Experimental Psychology: Applied*, 13: 224–248.

Kirkpatrick, D.L. (1976). Evaluation of training. In R.L. Craig (Ed.), *Training and Development Handbook: A Guide to Human Resource Development*, 2nd ed. New York: McGraw-Hill, 1–26.

Kraiger, K., Ford, J.K., and Salas, E. (1993). Application of cognitive, skill-based, and affective theories of learning outcomes to new methods of training evaluation. *Journal of Applied Psychology*, 78: 311–328.

Kraiger, K. (2002). Decision-based evaluation. In K. Kraiger (Ed.), *Creating, Implementing, and Managing Effective Training and Development: State-of-the-Art Lessons for Practice.* San Francisco: Jossey-Bass, 331–375.

Kripalani, S., Bengtzen, R., Henderson, L.E., and Jacobson, T.A. (2008). Clinical research in low-literacy populations: Using teach-back to assess comprehension of informed consent and privacy information. *IRB: Ethics & Human Research,* 30: 13–19.

Lehto, M.R. and Buck, J.R. (2008). *Introduction to Human Factors and Ergonomics for Engineers.* New York: Lawrence Erlbaum.

Lewis, J.R. (2006). Usability testing. In G. Salvendy (Ed.), *Handbook of Human Factors and Ergonomics,* 3rd ed. Hoboken, NJ: Wiley, 1275–1316.

Mayeaux, E., Jr., Murphy, P.W., Arnold, C., Davis, T.C., Jackson, R H., and Sentell, T. (1996). Improving patient education for patients with low literacy skills. *American Family Physician,* 53: 205–211.

Mayer, R., Hegarty, M., Mayer, S., and Campbell, J. (2005). When static media promote active learning: Annotated illustrations versus narrated animations in multimedia instruction. *Journal of Experimental Psychology: Applied,* 11: 256–265.

Mayer, R.E. and Moreno, R. (2003). Nine ways to reduce cognitive load in multimedia learning. *Educational Psychologist,* 38: 43–52.

Molenda, M. (2003). In search of the elusive ADDIE model. *Performance Improvement,* 42: 34–36.

Molenda, M., Pershing, J.A., and Reigeluth, C.M. (1996). Designing instructional systems. In R.L. Craig (Ed.), *The ASTD Training and Development Handbook,* 4th ed. New York: McGraw-Hill, pp. 266–293.

Moreno, R. (2006). Learning in high-tech and multimedia environments. *Current Directions in Psychological Science,* 15: 63–67.

Morrow, D., Leirer, V., Altieri, P., and Fitzsimmons, C. (1994). When expertise reduces age differences in performance. *Psychology and Aging,* 9: 134–148.

Morrow, D.G. and Rogers, W.A. (2008). Environmental support: An integrative framework. *Human Factors,* 50: 589–613.

Morrow, D.G., Hier, C.M., Menard, W.E., and Leirer, V.O. (1998). Icons improve older and younger adults' comprehension of medication information. *Journal of Gerontology: Psychological Sciences,* 53B: 240–254.

National Institute on Aging/National Institutes of Health (2007). Why population aging matters: A global perspective. Publication No. 07-6134. http://www.nia.nih.gov/sites/default/files/WPAM.pdf

Naumann, A.B., Wechsung, I., and Hurtienne, J. (2010). Multimodal interaction: A suitable strategy for including older users? *Interacting with Computers,* 22: 465–474.

Naveh-Benjamin, M. (2000). Adult age differences in memory performance: Tests of an associative deficit hypothesis. *Journal of Experimental Psychology: Learning, Memory, and Cognition,* 26: 1170–1187.

Newell, A., and Simon, H.A. (1972). *Human Problem Solving.* Englewood Cliffs, NJ: Prentice-Hall.

nwlink Website (2011). http://www.nwlink.com/~donclark/history_isd/addie.html

Oestermeier, U. and Hesse, F.W. (2000). Verbal and visual causal arguments. *Cognition,* 75: 65–104.

Oviatt, S. (2003). Multimodal interfaces. In J.A. Jacko and A. Sears (Eds.), *The Human-Computer Interaction Handbook: Fundamentals, Evolving Technologies and Emerging Applications.* Mahwah, NJ: Lawrence Erlbaum, 286–304.

Pak, R. and McLaughlin, A.C. (2010). *Designing Displays for Older Adults*. Boca Raton, FL: CRC Press.

Park, D.C., Smith, A.D., Lautenschlager, G., Earles, J.L. Frieske, D., Zwahr, M., and Gaines, C.L. (1996). Mediators of long-term memory performance across the life span. *Psychology and Aging*, 11: 621–637.

Péruch, P. and Wilson, P.N. (2004). Active versus passive learning and testing in a complex outside built environment. *Cognitive Processing*, 5: 218–227.

Pew Internet & American Life Project. 2004. *Older Americans and the Internet*. http://www.pewinternet.org/pdfs/PIP_Seniors_Online_2004.pdf

Pew Internet & American Life Project (2011a). Generations and cell phone ownership. http://www.pewinternet.org/Infographics/2011/Generations-and-cell-phones.aspx

Pew Internet & American Life Project (2011b). http://pewinternet.org/Trend-Data/Internet-Adoption.aspx

Purcell, K. (2011). Half of adult cell phone owners have apps on their phones. http://pewinternet.org/~/media//Files/Reports/2011/PIP_Apps-Update-2011.pdf

Purdie, N. and Boulton-Lewis, G. (2003). The learning needs of older adults. *Educational Gerontology*, 29: 129–149.

Reason, J. (1997). *Managing the Risks of Organizational Accidents*. Aldershot, Hampshire, UK: Ashgate.

Reigeluth, C.M. (2007). Order, first step to mastery: An introduction to sequencing in instructional design. In F.E. Ritter, J. Nerb, E. Lehtinen, and T.M. O'Shea (Eds.), *In Order to Learn: How the Sequence of Topics Influences Learning*. New York: Oxford University Press, 19–40.

Renkl, A. (2002). Learning from worked-out examples: Instructional explanations supplement self-explanations. *Learning and Instruction*, 12:, 529–556.

Renkl, A. and Atkinson, R.K. (2007). An example order for cognitive skill acquisition. In F.E. Ritter, J. Nerb, E. Lehtinen, and T.M. O'Shea (Eds.), *In Order to Learn: How the Sequence of Topics Influences Learning*. New York: Oxford University Press, 95–105.

Renkl, A., Stark, R., Gruber, H., and Mandl, H. (1998). Learning from worked-out examples: The effects of example variability and elicited self-explanations. *Contemporary Educational Psychology*, 23: 90–108.

Roscoe, S.N. (1971). Incremental transfer effectiveness. *Human Factors*, 13: 561–567.

Rose, F.D., Attree, E.A., Brooks, B.M., Parslow, D.M., and Penn, P.R. (2000). Training in virtual environments: Transfer to real world tasks and equivalence to real task training. *Ergonomics*, 43: 494–511.

Ruiz, J.G., Andrade, D.A., Freeman, D., and Roos, B.A. (2011). (personal communication). The virtual teach-back. *GRECC Laboratory of E-Learning and Multimedia Research*, Miami Veteran's Administration Medical Center.

Salas, E., Wilson, K.A., Priest, H.A., and Guthrie, J.W. (2006). Design, delivery, and evaluation of training systems. In G. Salvendy (Ed.), *Handbook of Human Factors and Ergonomics*, 3rd ed. Hoboken, NJ: John Wiley & Sons.472–512

Salden, R.J.C.M., Paas, F., Broers, N., and van Merriënboer, J.J.G. (2004). Mental effort and performance determinants for the dynamic selection of learning tasks in air traffic control training. *Instructional Science*, 32: 153–172.

Schaie, K.W. (1996). *Intellectual development in adulthood: The Seattle Longitudinal Study*. New York: Cambridge University Press.

Schillinger, D., Piette, J., Grumbach, K., Wang, F., Wilson, C., Daher, C., Leong-Grotz, K., Castro, C., and Bindman, A.B. (2003). Closing the loop: Physician communication with diabetic patients who have low health literacy. *Archives of Internal Medicine*, 163: 83–90.

Sharit, J., Czaja, S.J., Hernandez, M., Yang, Y., Perdomo, D., Lewis, J.L., Lee, C.C., and Nair, S. (2004). An evaluation of performance by older persons on a simulated telecommuting task. *Journal of Gerontology*, 59B: 305–316.

Sharit, J., Czaja, S.J., Hernandez, M.A., and Nair, S.N. (2009). The employability of older workers as teleworkers: An appraisal of issues and an empirical study. *Human Factors and Ergonomics in Manufacturing*, 19: 457–477.

Shepherd, A. (2001). *Hierarchical Task Analysis*. London: Taylor & Francis.

Smyer, M.A. and Pitt-Catsouphes, M. (2009). Collaborative work: What's age got to do with it? In S.J. Czaja and J. Sharit (Eds.), *Aging and Work: Issues and Implications in a Changing Landscape*. Baltimore: Johns Hopkins Press, 144–164.

Sorkin, D.H. and Heckhausen, H. (2006). Motivation. In R. Schulz (Ed.), *The Encyclopedia of Aging*, 4th ed. New York: Springer, 794–797.

Stoltz-Loike, M., Morrell, R.W., and Loike, J.D. (2005). Can e-learning be used as an effective training method for people over the age 50? A pilot study. *Journal of Gerontechnology*, 4: 101–113.

Sullivan, S. and Duplaga, E. (1997). Recruiting and retaining older workers for the new millennium. *Business Horizons*, 40: 65–69.

Sweller, J. (1994). Cognitive load theory, learning difficulty, and instructional design. *Learning and Instruction*, 4: 295–312.

Sweller, J. (2005). Implications of cognitive load theory for multimedia learning. In R.E. Mayer (Ed.), *Cambridge Handbook of Multimedia Learning*. New York: Cambridge University Press, 19–30.

Swezey, R.W. and Llaneras, R.E. (1997). Models in training and instruction. In G. Salvendy (Ed.), *Handbook of Human Factors and Ergonomics*, 2nd ed., New York: Wiley, 514–577.

Tennyson, R.D. (2010). Historical reflection on learning theories and instructional design, *Contemporary Instructional Technology*, 1: 1–16.

U.S. Census Bureau (2011). 90+ in the United States: 2006-2009. Report # ACS-17. http://www.census.gov/prod/2011pubs/acs-17.pdf

U.S. General Accounting Office (2001). Older workers: Demographic trends pose challenges for employers and workers. GAO-02-85. Available online http://www.doleta.gov/performance/guidance/gaoreports/d0285.pdf.

U.S. General Accounting Office (2003). *Older Workers: Policies of Other Nations to Increase Labor Force Participation*. http://www.gao.gov/new.items/d03307.pdf

Van Gerven, P.W.M., Paas, F., van Merriënboer, J.J.G., Hendriks, M., and Schmidt, H.G. (2003). The efficiency of multimedia learning into old age. *British Journal of Educational Psychology*, 73: 489–505.

van Merriënboer, J.J.G. (1997). *Training Complex Cognitive Skills: A Four-Component Instructional Design Model for Technical Training*. Englewood Cliffs, NJ: Educational Technology.

van Merriënboer, J.J.G., Clark, R.E., and De Croock, M.B.M. (2002). Blueprints for complex learning: The 4C/ID-model. *Educational Technology, Research and Development*, 50: 39–64.

Wankel, C. and Kingsley, J. (2009). *Higher Education in Virtual Worlds: Teaching and Learning in Second Life*. Bingley, UK: Emerald Group.

Wickens, C.D., Gordon Becker, S.E., Liu, Y., and Lee, J.D. (2004). *An Introduction to Human Factors Engineering,* 2nd ed. New York: Prentice-Hall.

Yerkes, R.M. and Dodson, J D. (1908). The relation of strength of stimulus to rapidity of habit-formation. *Journal of Comparative Neurology and Psychology,* 18: 459–482.

Index

Printed in the United States
by Baker & Taylor Publisher Services